AIR POLLUTION - MONITORING, QUANTIFICATION AND REMOVAL OF GASES AND PARTICLES

Edited by **Jorge Del Real Olvera**

Air Pollution - Monitoring, Quantification and Removal of Gases and Particles
http://dx.doi.org/10.5772/intechopen.73722
Edited by Jorge Del Real Olvera

Contributors

Patricia Córdoba Sola, Alireza Rahmati, Muhammad Zubair, Amina Adress, Rasa Zalakeviciute, Adrian Buenaño, David Sannino, Yves Philippe Rybarczyk, Joao Garcia, Bang Quoc Ho, H. N. Khue Vu, Li Ping Wong, Leonel Hernández Mena, Sergio Alonso Romero, Valeria Ojeda Castillo, Paz Elizabeth Alvarez Chavez, Jorge Del Real Olvera, Rosa Maria Ceron Breton, Julia Griselda Cerón Bretón, Gilma Gabriela Arenas Hernández, Evangelina Ramirez Lara, Atl Victor Cordova Quiroz, Manuel Muriel García, Jonathan Kahl, José Angel Solís Canul, Abril Rodríguez Guzmán, Alberto Espinosa Guzmán

First published in London, United Kingdom, 2019 by IntechOpen
IntechOpen is the global imprint of INTECHOPEN LIMITED, registered in England and Wales, registration number: 11086078, The Shard, 25th floor, 32 London Bridge Street
London, SE19SG – United Kingdom
Printed in Croatia

British Library Cataloguing-in-Publication Data
A catalogue record for this book is available from the British Library

Additional hard copies can be obtained from orders@intechopen.com

Air Pollution - Monitoring, Quantification and Removal of Gases and Particles, Edited by Jorge Del Real Olvera
p. cm.
Print ISBN 978-1-83880-193-9
Online ISBN 978-1-83880-194-6

For EU product safety concerns:
IN TECH d.o.o., Prolaz Marije Krucifikse Kozulić 3, 51000 Rijeka, Croatia,
info@intechopen.com or visit our website at intechopen.com.

Meet the editor

Jorge Del Real-Olvera PhD. is currently a Senior Researcher in the area of Environmental Technology at the Centre of Research and Assistance in Technology and Design of the State of Jalisco (CIATEJ) in Mexico. He completed his PhD studies in Environmental Science awarded by the Autonomous University of the State of Hidalgo. He conducted postdoctoral studies within the Electrochemistry group of the Autonomous Metropolitan University in environmental aspects. He is Project Leader in various topics of chemical engineering and environmental engineering such as Sustainable Drying of Agro-industrial Residues, Monitoring and Quantification of Air Pollution and Wastewater Treatment in diverse food industries. He has published several peer-reviewed journals, technical articles, chapters and books.

Contents

Preface IX

Chapter 1 **Air Emission Inventory 1**
Ho Quoc Bang and Vu Hoang Ngoc Khue

Chapter 2 **Airborne PM Impact on Health, Overview of Variables, and Key Factors to Decision Making in Air Quality 19**
João Nuno Pinto Miranda Garcia , Joana Borrega and Luís M. R. Coelho

Chapter 3 **Commuting on Public Transport: Health Risks and Responses 35**
Li Ping Wong, Haridah Alias, Nasrin Aghamohammadi, Nik Meriam Nik Sulaiman, Hualiang Lin and Jinshun Zhao

Chapter 4 **Air Pollution in an Urban Area of Mexico: Sources of Emission (Vehicular, Natural, Industrial, and Brick Production) 55**
Valeria Ojeda-Castillo, Sergio Alonso-Romero, Leonel Hernández-Mena, Paz Elizabeth Álvarez-Chávez and Jorge del Real-Olvera

Chapter 5 **Atmospheric N and S Deposition Fluxes in the Metropolitan Area of Monterrey, Mexico and Its Relation with Criteria Air Pollutants and Meteorological Conditions 75**
Rosa María Cerón Bretón, Julia Griselda Cerón Bretón, Jonathan Kahl, Evangelina Ramírez Lara, Atl Víctor Córdova Quiroz, Alberto Antonio Espinosa Guzmán, Manuel Muriel García, Gilma Gabriela Arenas Hernández, José Angel Solís Canul and Abril Rodríguez Guzmán

Chapter 6 **Urban Air Pollution Mapping and Traffic Intensity: Active Transport Application 99**
Rasa Zalakeviciute, Adrian Buenaño, David Sannino and Yves Rybarczyk

Chapter 7 **Dioxins and Furans: Emerging Contaminants of Air 111**
Muhammad Zubair and Amina Adrees

Chapter 8 **Emissions of Inorganic Trace Pollutants from Coal Power Generation 127**
Patricia Córdoba

Chapter 9 **Industrial Air Pollution Control 145**
Amir Hossein Matin and Ali Reza Rahmati

Preface

Currently, one of the most evident and dangerous contaminants aspects for the health of all living beings is air pollution. To understand the severity of this environmental problem, in this book the authors make an in-depth review of different environmental aspects on monitoring, quantification and elimination of emissions to the atmosphere, generated by the diverse anthropogenic activities in large cities. Contributors of this book have made an effort to put their ideas in simple terms without forgoing quality. The principal objective of this book is to present the most recent technical literature to all interested readers, particularly those who don't have access to expensive books and journals. Students, researchers and faculty members can employ this book for teaching purposes as well as for solving technical and industrial needs.

For a clearer representation about air pollution in big cities, in the beginning of every chapter there is a simple and convenient introduction to the basic principles and concepts handled in the area of emissions to the atmosphere from industrial and urban sources. In some chapters, the data obtained in the quantification and monitoring of gases and particles were employed in a mathematical simulation of the process.

Information about different topics in the field of air pollution is presented in this book. More specifically, the different methods to monitor of pollution in the atmosphere were analyzed; air quality modeling in indoor systems; inventory on emissions to air and new approaches to manage air quality and your effect on climatic change. This book covers the most recent concepts in the science of air pollutants and provides information on all the relevant components related to the problems of air pollution in urban areas and industrial zones.

I consider this book convenient for self-study by engineers and scientists in any discipline related to air pollution and also for all users who have some technical background. On the other hand, this book is meaningful to industrial and technical persons with some knowledge in the area of atmospheric pollutants. The chapters were elaborated to offer to the readers a quick practical overview of air pollution without entering into deep mathematical or theoretical considerations. I hope that this book will be useful to researchers working in the monitoring, control and depuration of air pollution.

Jorge del Real-Olvera, PhD
Environmental Technology, Centre of Research and
Assistance in Technology and Design of the State of Jalisco (CIATEJ)
México

Chapter 1

Air Emission Inventory

Ho Quoc Bang and Vu Hoang Ngoc Khue

Additional information is available at the end of the chapter

http://dx.doi.org/10.5772/intechopen.79928

Abstract

Emission inventory (EI) is a comprehensive listing of pollutants from all sources in a geographical area during a period of time. The development of EI database is very important to air quality management and helps us to find out main sources. In general, to develop a reliable EI for one specific area, the following steps are involved: (i) list the types of sources, (ii) determine the types of air pollutant emission from each of the listed sources, (iii) find out the emission factor (EF) for each of the concerned pollutants, (iv) determine the number and size of specific sources in the area, and (v) multiply the appropriate numbers from step (iii) and (iv) to obtain the total emissions and then sum up the similar emissions to obtain the total for the area. With emission inventory, we can define the main sources in the study area and have the best solution for clean air action. Besides, the result from EIs can be used to study in other sectors related to human health, economic loss from air pollution, etc. This chapter will describe the method for development of an air emission inventory for developing and developed countries.

Keywords: air emission inventory, Can Tho City, point sources, line sources, area sources, air quality management

1. Introduction

In general, air pollutants are defined as substances in the atmosphere, resulting from the activity of humans or from natural processes, causing harm to humans, plants, or animal lives, or damage to man-made materials and structures, or changes in the weather and climate, or interfering with the comfortable enjoyment of life or other human activities (Elsom, 1987).

There are six common air pollutants (criteria pollutants) which are commonly found in air anywhere. These main pollutants should be included in the emission inventory (EI): TSP

(PM_{10} and $PM_{2.5}$), NOx (the sum of NO and NO_2), SO_2, and CO. Other pollutants as VOC and CO_2 will be considered based on the characteristic of the study area and the aim of EI.

2. Design of an emission inventory

2.1. Defined sources

Normally, air emission sources are divided into three main categories: area source, point source, and line source. Each source includes many subcategories based on its distributed characteristic, which was shown in detail in **Figure 1**.

2.1.1. Point sources

Point sources are stationary sources that are typically large manufacturing or production plants that include confined, chimney or stack emission points. Within a given point source, there may be several units of combustion, such as several boilers. The point sources can be hospital incinerators, hospital boilers, hotel boilers, crematorium, and industries.

2.1.2. Area sources

Area sources are defined as sources that are too small and/or too numerous to be considered as point sources (US EPA, 1996). In aggregate, these sources may contribute a significant proportion of the total airshed emissions. Area sources include a wide range of sources, such as residential cooking facilities, rice straw/biomass burning, gas stations, construction activities, printing facilities, auto repair facilities, paint spraying facilities, and so on. They are presented as an area on the map.

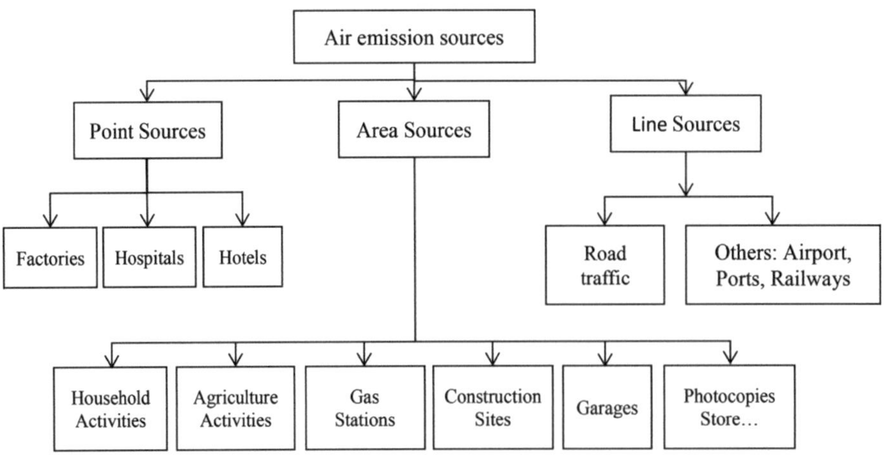

Figure 1. Air emission sources.

2.1.3. *Line sources (mobile sources)*

Line sources, also considered as mobile sources, are vehicles, engines, and equipment that generate air pollution and that move or can be moved from place to place. Mobile sources are classified under two smaller sources: on-road and non-road sources. On-road sources include vehicles used on roads for transportation of passengers or freight, such as passenger cars, trucks, and motorcycles that may be fueled with gasoline, diesel fuel, or alternative fuels, such as alcohol or natural gas. Non-road sources include gasoline and diesel powered vehicles, engines, and equipment used for construction, agriculture, transportation, recreation, and many other purposes. For example, non-road sources include marine vessels, aircrafts, railways, tractors, etc. They are presented as line shape on the map.

2.2. Emission inventory methods

2.2.1. *Point sources*

The aim of point source emission inventory is to give a comprehensive table sheet of sources, their characteristics, and annual emissions. A source which is large enough and contributes a huge amount of emissions can be considered as a point source, such as the hospital boilers, crematoriums, factory's chimneys, hotel boilers, etc. So the first step is to identify all sources in the study area and classify them into categories based on product characterization and combustion process. The list of these sources and emission data can be provided from local Province Officer and Department of Industry. For estimating emission, further process information from each source type needs to be provided directly from questionnaire or inspection. These data are specific for each category and also used for defining activity data, air emission process, and for input of dispersion model. The data normally require: factory type, size, geographical position (coordinates), number of employees, type of industrial activity: what is produced, combustion activities characterized by firing configuration, operating conditions, fuel specification, fuel consumption, raw material used in industrial/manufacturing process, storage of raw material, equipment age, height of stacks (and if available: inner diameter, volume flow in m^3/operating hour), hours of operation per year/season, air pollution abatement equipment, and continuous emission monitoring.

Example questionnaire for point source is shown in Appendix A

The emissions for each source category are calculated using Eq. (1) and emission factors. Due to the lack of data monitoring from factories, emission factors are adopted. They can be taken from these following sources, appropriate for production type and process:

- EMEP/EEA Emission Inventory Guidebook (CORINAIR): this technical guidance is for preparation of national emission inventories. The joint European Monitoring and Evaluation Programme/European Environment Agency (EMEP/EEA) air pollutant emission inventory guidebook supports the reporting of emission data under the United Nations Economic Commission for Europe (UNECE) Convention on Long-range trans boundary Air Pollution (CLRTAP) and the European Union National Emission Ceilings Directive. It provides expert guidance on how to compile an atmospheric emission inventory. The guidebook is published by the EEA with the CLRTAP Task Force on Emission Inventories and Projections responsible for the technical content of the chapters. This guidebook was most recently revised in December 2013, providing a completed emission inventory process. It also contains emission factors for many pollutants, as well as information on emissions from natural and organic sources. It can be accessed at http://www.eea.europa.eu//publications/emep-eea-guidebook-2013.

- United States Environmental Protection Agency—Compilation of Air Pollutant Emission Factors (US EPA—AP 42): originally serving the medical public in the United States (US) in 1968, this includes many basic emission factors. The most recent edition was released in January 1995 and has been supplemented since then. Since December 2005, US EPA has given users the opportunity to search and use the latest emission factors on the internet, supporting incorporation of almost any source into EIs. This allows the user to search for the latest and best emission factors for most point and area source categories, including a rating on the quality of the particular emission factor. It can be accessed at http://cfpub.epa.gov/oarweb/index.cfm?action=fire.main.
- Intergovernmental Panel on Climate Change (IPCC) 2006—Greenhouse Gases Emission Inventory Guidebook: This guides people to estimate greenhouse gas (GHG) emissions as CO_2 equivalent. IPCC can be used to take emission factors from other combustion activities [2]. The information can be accessed online. The emission factor, which is taken from this guide, should be explained and clearly stated in the report.

2.2.2. Area sources

Area sources are similar to point sources, including small sources with insignificant emissions, evenly distributed in a wide range area. The aggregation of all these sources contributes a considerable emission to air quality. For distributing on the map, emissions from these sources are estimated by each grid cell.

In the beginning, a list of all sources and their information are collected from related local Province Office or other related Departments, as well as the data of inhabitants, the number of households, and the density rate in each ward. These data are used for calculating emission from common sources distributed evenly, such as household cooking, using a top-down approach by which the total emissions for the whole region are calculated, and then they are distributed by each grid cell according to the number of households and the density rate. Sources with the same characteristic are classified under a category, and an appropriate emission estimation method is chosen depending on the source characteristic and the data available. The total emissions of one grid include the emission from household activities and from other sources.

Source categories are different depending on the feature of the study area, and some common categories for cities in developing countries are: biomass (straw) burning, solid waste open burning, household cooking, smoke for meat and fish curing, gas stations, construction, concrete mixing, garages, commercial/consumer solvent used, and graphic arts.

Normally, within one source category (k), combustion technology and fuel type used are similar so that generalized emission factor and fuel consumption rate (or number of source category k) can be used for estimating emission as in Eq. (1). So the main data collected from survey is fuel consumptions, fuel type, and source type that are included in questionnaires for area sources shown in Appendix A. Though several source categories such as photocopies and gas station emissions are not from combustion but from evaporative VOCs, they are also calculated by using this equation with appropriated emission factors.

$$\mathbf{E}_{\mathbf{k,i}} = \mathbf{A}_{\mathbf{k}} \times \mathrm{EF}_{\mathrm{k,i}} \tag{1}$$

where

$E_{k.i}$ = total emission from pollutant i in source category k in selected grid cell,

A_k = activity rate: that is, the amount of fuel burnt of source category k in the selected grid cell or surrogate data, and

$EF_{k,i}$ = appropriate average emission factor for pollutant i for source category k.

Emission factors for these sources are taken from USEPA, EMEP/EEA Emission Inventory Guidebook or from studies in the local university.

2.2.3. *Line sources (mobile sources)*

Line (mobile) sources include on-road sources and non-road sources. On-road sources are the emission from traffic which is distributed all around the area, directly affect human health. It is also the main contributor for this source type due to the huge number of vehicles. Non-road sources are emission from harbors, airports, and railways.

The emission from mobile sources can be estimated based on the fuel consumption in the study area and the average emission factor, with Eq. (2). Vehicle categories to be considered are motorcycles, passenger cars, light duty vehicles, heavy duty vehicles, and busses/coaches in which fuel includes gasoline, diesel, liquid petroleum gas (LPG), and natural gas if available. This method is a top-down approach which is suitable for estimating emission in a huge area. However, top-down approach is high level of uncertainty.

$$E = \sum_{i=1}^{n} (F\,C_i \times EF) \tag{2}$$

where

E: total emissions (include emission from each pollutant from each vehicle category and fuel type),

FC_i: fuel consumption (for each vehicle category and each fuel type), and

EF: fuel consumption—specific emission factor (for each pollutant and vehicle category and fuel).

For a more detailed approach, subdivided emission into hot emission, cold-start emission, and extra emission as in Eq. (3) can be applied. Generally, the total emissions are the sum of hot emissions (the emissions when a vehicle is running, with the engine thermally stabilized), cold emissions (the emissions when a vehicle is running, during engine warm-up), and evaporative emissions (the emissions from fuel due to the difference between the temperature of engine and ambient temperature). The equation for total emissions is shown below:

$$E_{total} = E_{hot} + E_{cold} + E_{extra} \tag{3}$$

where

E_{total}: total emissions of pollutant i,

E_{hot}: hot emissions,

E_{cold}: cold emissions, and

E_{extra}: evaporation emissions, tyre, and brake wear.

Hot emissions of pollutant i are calculated by the equation below:

$$E_{(hot,i,k,T)} = N_k \times L_{(a,T)} \times e_{(h,i,k,T)} \tag{4}$$

where

$E_{(hot,i,k,T)}$: hot exhausted emissions of pollutant i, produced in the relevant period by vehicles of technology k driven according to traffic situation T (g),

N_k: number of vehicles (veh) of technology k in operation in the period concerned,

$L_{(a,T)}$: length of street section a (in km) with traffic situation T (km), and

$e_{(hot,i,k,T)}$: emission factor in (g/km) for pollutant i, relevant to the vehicle technology k, operated in traffic situation T.

Cold-start emissions are calculated as extra emissions and usually have a small amount compared with hot emissions. Especially in tropical countries with high temperature all year, cold-start emissions are negligible. Eq. (5) is used to calculate cold-start emissions.

$$E_{(cold,i,i)} = \beta_{i,k} \times N_k \times M_k \times e_{hot;i,k} \times \left(e^{cold}/e^{hot}\big|_{i,k} - 1\right) \tag{5}$$

where

$E_{(cold,i,i)}$: cold-start emissions of pollutant i, produced by vehicles category j and technology k,

$\beta_{i,\,k}$: fraction of mileage driven with a cold engine or the catalyst operated below the light-off temperature for pollutant i and vehicle technology k. β depends upon the ambient temperature,

N_k: number of vehicles of technology k in circulation,

M_k: total mileage per vehicle in vehicle technology k, and

$e^{cold}/e^{hot}|_{i,k}$: cold/hot emission quotient for pollutant i and vehicle technology k.

Evaporation emissions generated by gasoline vehicle are the result of the increase in ambient temperature during the daylight hours and the expansion of the fuel and vapor in the tank. Evaporation emissions are estimated as in Eq. (6).

$$E_{VOC} = \sum_j N_j \times EF_{VOC,j} \times 365 \tag{6}$$

where

E_{VOC}: emission of VOC (g/year),

N_j: number of vehicles in category j,

$EF_{VOC,j}$: emission factor of VOC for vehicle in category j (g/veh/day),

j: vehicle category, and

365: number of days per year.

Emissions from tyre and brake wear also considered estimating since they contribute a significant amount of particulate matter (PM), especially those from motorcycles. These are considered as extra emissions and causes from friction between the tyre and the road surface when the vehicle is running. There are many calculation methods for this estimation since it is not complicated and requires simply data, for example, the estimation using Tier 1 emission factors from CORINAIR requires the data of the kilometers the vehicle which has annually traveled in the study area, as in Eq. (7).

$$TE = \sum_{j=1}^{n} \left(N_j \times M \times EF\right) \tag{7}$$

where

TE: total emissions of PM for the defined time period and spatial boundary (g),

N: number of vehicle in category j within the define time period,

M: average mileage driven per vehicle in category j during the defined time period (km), and

EF: mass emission factor for pollutant i and vehicle category j (g/km).

Emission factors can be found in emission factor handbooks and research papers such as CORINAIR, EMEP, AP-42, etc. EFs of pollutants depend not only on the generation processes but also on pollution control systems that have been applied. Most of the traffic emission models are intergraded emission factor database with clear reference, and it can also be modified by the user in accordance with the study area.

In general, for urban traffic emission estimation, an emission model is adopted since it can handle a lot of complicated data and consider different fleet compositions and driving conditions in different streets and integrated emission factor database. The advantages of emission models are the time saved, the availability of many estimating options, and the reduction of the calculating uncertainty. There are several emission models available, like COPERT,

MOBILE-5, MOVES, IVE, MOBILEV, EMISENS, etc., but not all emission models are appropriate for calculating traffic emission in single street levels with different traffic situations. Most models were developed for calculations on national or regional level and use, therefore, more aggregated emission factors. An emission model used for an emission inventory on an urban scale with high spatial resolution should fulfill at least the following criteria: highly differentiated emission factor database, a broad selection of street types and traffic situations, an internal module to calculate cold start emissions, a representative and appropriate fleet composition by subsegments for each vehicle category, appropriate default fuel quality parameters, a module to calculate fuel consumption caused by air condition, access to default values in case adoptions are necessary, a user-friendly interface, a user-friendly data input tool, and a comprehensive description of the model.

Input data for emission models usually require the following information which is collected from surveys, previous researches, analysis and observation from the study area: defining main street network/defining road categories, length and width of road, number of lanes, speed limit for each road type/category, define vehicle type/category, average daily traffic, vehicle fleet/composition, vehicle registration database, and vehicle engine type, capacity, emission stage, fuel use, weight, and vehicle traffic curve.

2.2.4. Non-road sources

Non-road sources include gasoline and diesel powered vehicles, engines, and equipment used for construction, agriculture, transportation, recreation, and many other purposes. For example, non-road sources include marine vessels, aircrafts, railways, tractors, etc.

2.2.5. Emission from seaport

There are many different approaches to develop an emission inventory for ports. They vary significantly according to the availability of time, money, and effort. For example, a detailed inventory approach requires detailed data on vessels and land-based equipment characteristics and activities, as well as detailed information on port geography and ship paths within the port. This is the best practice for all ports; however, its applications are limited with available resources. Meanwhile, the streamlined inventory approach requires less input data. The methodology can be tailored to the amount of data available. In some cases, the emission inventory can be developed using extrapolation data.

To calculate air emissions in the port, a streamlined air emission inventory approach was used following US EPA guidelines [3].

2.2.5.1. Ocean going vessels (OGVs)

The available data for some ports in developing countries are not sufficient to develop a detailed inventory by calculating emissions from each vessel. The US EPA guidance [3] proposes a mid-tier approach for calculating air emissions in ports and allows us to calculate emissions from each category using a single representative vessel specification and operating pattern, rather than calculating emissions from every different vessel call. In this case, a detailed or mid-tier approach was selected based on the available data.

Emissions were then calculated using Eq. (8):

$$E = P \times LF \times A \times EF \tag{8}$$

where

E = emissions (g),

P = maximum continuous power rating (kW),

LF = load factor (%),

A = activity (hours), and

EF = emission factor (g/kWh). Emission factors were taken from US EPA [3].

Main engine load factors were calculated using Eq. (9):

$$LF = (AS/MS)^3 \tag{9}$$

where

LF = load factor (%),

AS = actual speed (knots), and

MS = maximum speed (knots).

2.2.5.2. *Cargo-handling equipment (CHE)*

Emissions were calculated for each category using Eq. (10):

$$E = N \times P \times LF \times A \times EF \tag{10}$$

where

E = emissions (g),

N = number of items,

P = maximum continuous power rating (kW),

LF = load factor (%),

A = activity (hours), and

EF = emission factor (g/kWh).

2.2.6. *Emission from airport*

Emissions from this type of sources are often overlooked since it is not a typical source. For estimating emission, the following data are required:

- Fuel consumption for each flight phases (landing and taking-off (LTO), cruise phase…)
- Emission factor for each flight phases and aircraft type

- Type of aircraft available
- Number of landing and taking off for domestic aviation and international aviation per aircraft type

Basically, emissions are calculated from aircraft type, fuel type used, and emission factor related. Based on the data available, a simple or detail approach can be applied. Typically, Tier 2 approach from EMEP/EEA emission inventory guidebook 2013 is applied as in Eq. (11) since there are no sufficient data for higher tier. Related emission factors are also taken from this document.

$$E_{pollutant} = \sum_{aircraft\ types} \left(AR_{fuel\ consumption,aircraft\ type} \times EF_{pollutant,aircraft\ type}\right) \quad (11)$$

where

$E_{pollutant}$: annual emission of pollutant for each of the LTO and cruise phases of domestic and international flights,

$AR_{fuel\ consumption,\ aircraft\ type}$: activity rate by fuel consumption for each of the flight phases and trip types, for each aircraft type, and

$EF_{pollutant,\ aircraft\ type}$: emission factor of pollutant for the respective flight phase and trip type, for each aircraft type.

Calculation steps for this approach are as the following:

- Obtain the total amount of fuel sold for all aviation (in ktonnes).
- Obtain the total amount of fuel used for domestic aviation (in ktonnes).
- Calculate the amount of fuel used for international aviation by subtracting the domestic aviation (step 2) from the total fuel sold (step 1) (in ktonnes).
- Obtain the total number of landing and taking off carried out per aircraft type for domestic aviation.
- Calculate the fuel use for landing and taking off activities per aircraft type for domestic aviation.
- Calculate the total fuel use for domestic cruise by subtracting the total amount of fuel for LTO activities found in step 6 from the total in step 2.
- Estimate the emissions from domestic LTO activities per aircraft type. The number of landing and taking off for each aircraft type is multiplied by the emission factor related to the particular aircraft type and pollutant.
- Estimate the emission from domestic cruise activities. Use the domestic cruise fuel use and the corresponding emission factor for the most common aircraft type used for domestic cruise activities.
- Calculate the total emission for landing and taking off activities for domestic aviation.

- Calculate the total emission for cruise activities for domestic aviation.
- Do the same for calculating for international aviation.

Data on the aircraft types and the number of flights were collected directly from the airport and from the Department of Transportation (DOT).

2.2.7. Emission from railway

Emissions from railway are mainly from locomotives which generally are one of the three types: diesel, electric, or steam. Like other sources, emissions are the product of fuel combustion with CO, NOx, SO_2, PM, hydrocarbons, and other heavy metals. According to EMEP/EEA, the locomotives fall into three categories: shunting locomotives—used for shunting wagons whose power output is typically in the range of 200–2000 kW, railcars—used for short distance rail traction with power output about 150–1000 kW, and line haul locomotives—used for long distance rail traction both for freight and passengers with power output around 400–4000 kW.

The methodology to estimate the emission for railway is from the total fuel type used and related emission factors as in Eq. (12) as follows:

$$E_i = \sum_m \sum_j \left(FC_{j.m} \times EF_{i.j.m}\right) \quad (12)$$

where

E_i: emission of pollutant i for the period concerned in the inventory (kg or g),

$FC_{j.m}$: fuel consumption of fuel type m used by category j for the period and area considered (tonnes),

$EF_{i.j.m}$: emission factor of pollutant i for each unit of fuel type m used by category j (kg/tonnes),

m: fuel type (diesel and gas oil), and

j: locomotive category (shunting, rail-car, and line haul).

2.3. Case study

2.3.1. Can Tho City

2.3.1.1. Overview

In this section, the process to apply the above theory for conducting an air emission inventory Can Tho City (CTC) is described. The emission inventory includes basic information about the dispersion of air pollutants, such as location, sources, and emission rates. The emission inventory assessment is necessary because it helps businesses implement effective policies to control their emissions, while it also helps the administration assess the current emissions and construct air quality management strategies in the future.

Can Tho is the fourth largest city in Vietnam and the largest city in the Mekong Delta. It is noted for its floating market, rice paper-making village, and picturesque rural canals. In 2013, in an area of 140,894.9 ha, the city had 1,232,260 inhabitants. Can Tho city is located on the south bank of the Hau River, a branch of the Mekong River. The city is located at N 9055′08″–10,019′38″; E 105013′38″–105,050′35″. It consists of five urban districts (Thot Not, O Mon, Binh Thuy, Ninh Kieu, and Cai Rang) and four suburban districts (Vinh Thanh, Thoi Lai, Co Do, and Phong Dien). Can Tho City was administratively formed on 1st January, 2004, when the old province of Can Tho was divided into Hau Giang province and Can Tho City; Can Tho City then directly belongs to the Central Committee (City Profile, 2013).

In general, there are four main types of emission sources: point, area, and on-road, and non-road mobile sources. **Figure 2** shows the locations of emission sources in the study domain. Different symbols represent various kinds of point and area sources including gas stations, garages, construction sites, industrial factories, hospitals, hotels, etc.

2.3.1.2. Collecting data and calculating

EI was conducted for two districts: Ninh Kieu and Binh Thuy (NKBT) (**Table 1**).

2.3.1.2.1. Total emission from area sources

The highest area emission source type for all pollutants in NKBT is burning fuel activities from households/residential activities (**Table 2**).

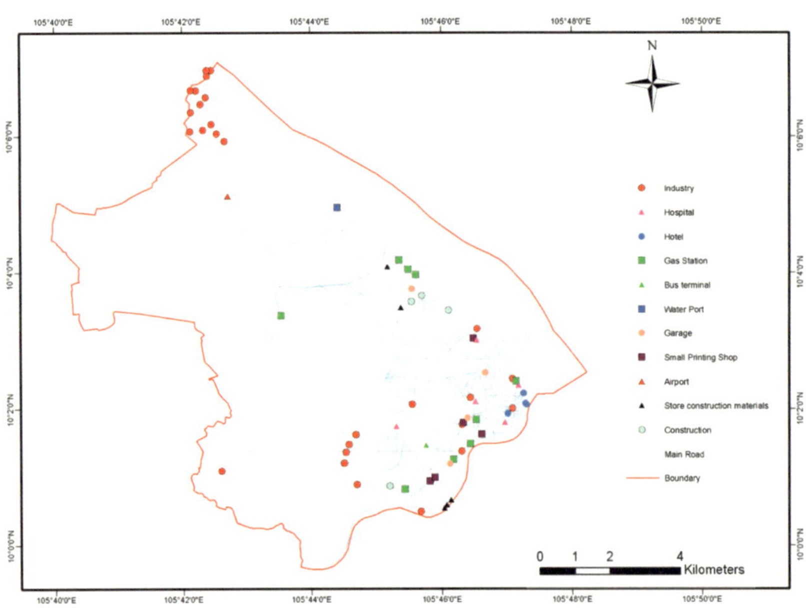

Figure 2. The locations of point, mobile, and some area emission sources that have been visited/interviewed by the data collection team.

Type of source	Pollutants (tonne/year)					
	NO_X	CO	SO_2	VOC	CO_2	PM_{10}
Construction sites	0.08	0.02	0.00	0.01	—	6.93
Photocopy stores	—	—	—	0.84	—	—
Construction material vendors	—	—	—	—	—	8.73
Gas stations	—	—	—	0.11	—	—
Restaurants—Cooking	0.62	23.98	3.65	2.92	695.61	2.33
Burning rice straw	8.39	342.39	0.66	25.77	4333.26	33.50
Households	77.28	2678.05	161.50	429.82	88,162.37	328.77
Car garages	—	—	—	1.41	—	—
Total	**86.37**	**3044.44**	**165.81**	**460.88**	**93,191.24**	**380.26**

Table 1. Total emissions from area sources in NKBT.

Activity	Emission rate (tonne/year)					
	NO_X	CO	SO_2	VOC	CO_2	PM_{10}
Hotels	47.35	9.71	3.21	5.90	16,560	1.46
Hospital	3.41	0.44	0.31	0.17	492.52	0.13
Industry	1006	2620	30.67	795.25	643,070	8.56
Total	1057	2630	34.19	801.32	660,123	10.15

Table 2. Total emissions for point source in CTC.

2.3.1.2.2. Point source

Point sources in CTC also include emissions from industries, hospital incinerators, and hotel boilers. The total emission for this source is presented in **Table 3**.

Air pollution emissions that are NO_X, SO_2, CO, CO_2, and VOC (NMVOC) from industrial activities are the largest among the three categories of point source emitters.

2.3.1.2.3. Mobile source

2.3.1.2.3.1. On-road mobile source

EI for mobile sources in CTC was implemented using Mobilev model, as in the case of Bac Ninh City. It shows that the main pollutant from this source is CO with the main contributor, motorcycle. The emission from side street generated from small street in residential (include alleyways and passageways) also contributes a significant amount of emissions.

Source type	CO	CO_2	HC	NOx	PM	SO_2
Motor	5843.931	47,371.67	495.577	132.99	75.835[a]	15.936
Car	146.33	7481.723	26.531	25.827	7.011	2.619
Bus	10.168	3572.496	3.715	42.449	1.547	1.262
LDV	63.97	1926.505	7.945	8.623	1.621	0.677
HDV	22.112	6782.04	8.23	82.085	3.33	2.396
Bus terminal	0.72	300.85	0.17[b]	2.56	0.11	0.05
Side streets	6964.53	75,908	923.82	421.91	97.49	23.66
Tyre and brake wear	—	—	—		80.671	—
Diurnal evaporation	—	—	1035.46[c]	—	—	—
Total	**13,051.76**	**143,343.28**	**2501.45**	**716.44**	**267.62**	**46.60**

Note: [a] is emissions of PM_{10}; [b] is emissions of NMVOC; [c] is VOC; and HC is hydrocarbon.

Table 3. Total emissions from on-road sources (tonne/year).

2.3.1.2.3.2. Non-road mobile source

2.3.1.2.3.2.1. Emissions from Hoang Dieu Port

We conducted a survey with an emission declaration table. For ocean going vessels: in general, anchorage areas for OGVs meet the standard, and the activity data are used in calculation including: name of OGVs, arrived day, leaving day, good categories (container, roll-on/roll-off, package, etc.), gross register tonnage (GRT), dead weight tonnage (DWT), power, etc. Some information included: time and speed in reduced speed zone (RSZ) (transit) mode, in maneuver mode, etc.

For CHE, there are details about the number, type, capacity, and equipment life, allowing the preparation of a basic inventory.

Emissions from Hoang Dieu Port in 2014 are from 22 OGVs and 11 CHE (including cranes and forklift trucks) shown in **Table 4**.

As you can see, the total emissions are mainly generated from CHE, except for SO_2 from OGVs. This is the result of using fuel with a high content of sulfur. 73.3% of PM_{10}, 74.4% of $PM_{2.5}$, and 77.3% of CO_2 come from CHE.

2.3.1.2.3.2.2. Emissions from Can Tho airport

Can Tho International Airport (CTIA) is located in Binh Thuy district, and emissions from this source can affect people living in CTC. The numbers of flights from CTIA are shown in **Table 5**.

Emission factors for air plane are referred in the documents:

***EMEP/EEA Emission Inventory Guidebook 2009, Section 2, **Tables 3–5**, page 8.*

Source	NO_X	PM_{10}	$PM_{2.5}$	VOC	CO	SO_2	CO_2
OGV	0.97	0.12	0.11	0.04	0.09	0.96	182.01
CHE	9.82	0.33	0.32	0.80	3.16	0.13	618.28
Total	**10.79**	**0.45**	**0.43**	**0.84**	**3.25**	**1.09**	**800.29**

Table 4. Total emissions from OGVs and CHE in 2014 (tonne/year).

Type of plane	Number of flights		Total flight (flight/year)
	Domestic (flight/year)	International (flight/year)	
A321	1460	10	1470
A320	450	0	450
AT72	730	0	730
Total			**2650**

Table 5. Number of flights in CTIA (2014).

S. Caserini. M. Monguzzi. A. Fraccaroli. M. Moretti. E. Angelino, Emission inventory in Lombardy Airport (Italy): Methodology and result,* ***Table 4**, page 8.*

Emissions from airports can be generated through airplanes taking off and landing. Air emissions from this source are estimated by using the number of flights per year and emission factors related (kg/flight) for each type of airplane. **Table 6**, which contains the emission from airport, shows that the emission rates of NO_x, SO_2, CO, CO_2, VOC, and PM_{10} are 14.41, 1.02, 23.79, 3048, 2.31, and 0.22 tonne/year, respectively.

2.3.1.2.3.3. Total non-road mobile emissions

Emissions of CO, VOC, and CO_2 from the airport are much higher than the marine port within the non-road mobile emission source category (**Table 6**).

2.3.1.2.3.4. Total emissions in CTC

Air emissions in NKBT, Can Tho City, are dominated by emissions from on-road mobile sources (mainly from motorcycles) for CO and VOC, by area sources (mainly from household energy consumption) for SO_2 and PM10, and dominated by point sources (mainly from

Pollutant	NO_X	CO	SO_2	VOC	CO_2	PM_{10}
Airport	14.41	23.788	1.021	2.308	3048.369	0.224
Marine port	10.790	3.243	1.088	0.838	800.284	0.452
Total	**25.200**	**27.031**	**2.109**	**3.146**	**3848.653**	**0.676**

Table 6. Total emissions from non-road mobile sources in NKBT (tonne/year).

Source	Emission parameters (tonne/year)					
	NO_x	CO	SO_2	VOC	CO_2	PM_{10}*
On-road mobile sources	716.45	12,173.02	45.78	2178.91	134,268.86	183.52
Non-road mobile sources	27.76	27.75	2.16	3.32	4149.50	0.79
Point sources	1057.47	2630.90	34.19	801.32	660,124.38	10.14
Area sources	86.38	3044.44	165.82	460.88	93,191.24	380.26
Total	**1888.06**	**17,876.11**	**247.95**	**3444.43**	**891,733.98**	**574.71**

*Note: * Only emission from on-road mobile source is for PM, and in* **Table 7** *considers it as* PM_{10}.

Table 7. Total emissions from all sources in NKBT in 2014 (ton/year).

industrial activities) for NO_x and CO_2. It was found that on-road traffic emitted a significant amount of pollutants into the atmosphere. Also, household activities, which were classified as area sources, played an important role [1].

Point sources are the biggest emitters of NO_x and CO_2; area sources are the main emitters of SO_2 and PM_{10}; and on-road mobile sources are the main emitters of CO and VOC (**Table 7**).

3. Conclusions

Emission inventory is a basic step of air quality management. There are several approaches of EI as described above, and based on each study area, considered using EI approaches consistent with the available data. High accuracy in EI is proportional to time consuming and costly and inversely. In general, less time and low cost are proportional with inaccuracy. And in general, an EI should be done annually or at least every 5 years to update the air emission status and to observe the trend of major air pollution sources and evaluate the effectiveness of air quality control.

A. Restaurants/households/street cookings

1. General Information:

Restaurant/Households-Residential/Street cooking name: ....

Address:..

Phone: ....

Fax: ....

Coordination:...N...E

2. Fuel use:

2.1 LPG

Size. Capacity..kg

Amount of use ....day/tank.

Operating period..hour/day.

Period of time: ....hour/day.

2.2. Charcoal:

Types:

Amount of use...kg (day/month/year)

Operating period..hour/day.

Period of time...hour/day.

2.3. Elictric stove/ Microwave

Electric power...W

Amount of use...day.

2.4. Other types

Day……month….year…

Informant Interviewer

Author details

Ho Quoc Bang* and Vu Hoang Ngoc Khue

*Address all correspondence to: bangquoc@yahoo.com

Air Pollution and Climate Change Department, Institute of Environment and Resources (IER), Viet Nam National University Ho Chi Minh City (VNUHCM), Vietnam

References

[1] Bang QH, Khue VHN, Tam NT, Lasko K. Air Pollution Emission Inventory and Air Quality Modeling. Can Tho City, Mekong Delta, Vietnam: Air Quality, Atmosphere & Health; 2017. pp. 1-13

[2] IPCC. IPCC Guidelines for National Greenhouse Gas Inventories. 2006. Available from: http://www.ipcc-nggip.iges.or.jp/public/2006gl/vol4.html

[3] US EPA. Current Methodologies in Preparing Mobile Source Port—Related Emission Inventories; 2009

Chapter 2

Airborne PM Impact on Health, Overview of Variables, and Key Factors to Decision Making in Air Quality

João Nuno Pinto Miranda Garcia ,
Joana Borrega and Luís M. R. Coelho

Additional information is available at the end of the chapter

http://dx.doi.org/10.5772/intechopen.79695

Abstract

This chapter intends to contribute to the understanding of the multiple aspects related to particulate matter (PM) in an air urban environment, in particular, regarding its impact on human health. A general overview of variables and key factors is presented to identify, relate, and understand the diverse and multidisciplinary variables that contribute to PM concentration in urban environments associated with health impacts. This relation is difficult to quantify, given the numerous variables that are interlinked due to the multidisciplinary aspects involved. Our aim is to identify the main multidisciplinary aspects, namely, meteorology, urban geometry, buildings, roads and footpaths, road traffic, industries, air concentration measurements, and health. The main strategic aspects for decision making related to airborne PM impact on health are also discussed.

Keywords: airborne PM, urban air quality, variables, key factors, air pollution, public health

1. Introduction

According to the World Health Organisation (WHO), air pollution is defined as 'The presence of one or more contaminants in the atmosphere, such as dust, fumes, gases, gas, 'fog', odour or vapour in quantities or with characteristics, and of a duration that may be detrimental to human, animal or plant life, to property or that interferes unfavourably in the comfortable enjoyment of life or property' [1]. Thus, it is said that an atmospheric condition, where there are substances at higher concentrations than ambient levels, is polluting if there are effects that can be measured on humans, animals, flora or materials. Within air pollutants, particulate matter (PM) is undoubtedly one of the most important in terms of impact on health. Indeed, all over the years, a growing number of scientific studies have attempted to correlate the outcome of possible adverse health effects with exposure to levels of particle concentration in atmospheric air. In order to try to understand and analyse this correlation, it is necessary to understand what particles are in the atmospheric air, their constitution, their origin, and the mechanisms that govern them. Generally, a very large group of pollutants in the air are grouped together under the designation of particles and may originate from such sources as cars, steelworks, thermal power plants, heating systems, cement plants, volcanoes, deserts, and oceans. Generally speaking, the NIST [2] definition for a particle is 'any condensed-phase three-dimensional discontinuity in a dispersed system may be considered a particle'. However, in terms of atmospheric pollution, a particle can be defined as a solid, liquid or solid and liquid dispersed matter, whose individual aggregates are larger than small molecules in diameter greater than 0.0002 μm but less than 500 μm.

2. Atmospheric pollutants

Atmospheric pollutants can be classified according to their origin and can be classified as: primary, if they are emitted directly by identifiable sources (fixed, mobile, and natural sources), for example: NO_X, SO_X, CO, particulates, hydrocarbons, and metals; secondary, if they are generated in the atmosphere by the reaction between two or more primary pollutants, or by the reaction between the current air constituents, and by photoactivation, hydrolysis or oxidation, for example: O_3, other photochemical oxidants such as peroxyacetyl nitrate and oxidised hydrocarbons (HC). They may also be classified by their chemical composition, being divided by: organic, which include hydrocarbons (HC), alcohols, and esters; inorganic, which include NO_X, SO_X, CO and CO_2, and metals. They can also be classified according to their physical state, which can be: gases (for example: NO_X, SO_X, CO, and CO_2), O_3; particles (solid or liquid), which are usually identified as dispersed material, for example: metal particles, asbestos, carbon (C), resin, nitrate, sulphate, bacteria, dioxins, and furans. There are certain contaminants that normally exist in all urban areas, which are referred to as reference pollutants (CO, NO_X, O_3, SO_2, and PM_{10}). Their concentrations vary and depend on the level of industrial activity and the traffic. In addition to these, hundreds of other compounds specific to each type of industry exist, for example: acrylonitrile, benzene, dichloromethane, formaldehyde, polycyclic aromatic hydrocarbons (PAHs), vinyl chloride, carbon disulfide, 1,2-dichloromethane, styrene, tetrachlorethylene, toluene, trichlorethylene, arsenic, asbestos, and heavy metals.

3. Relation between meteorology and atmospheric pollution

Atmospheric conditions vary from day to day in all locations. The state of the weather at a given location and at a given time is characterised by the cloud cover and its type, the intensity and direction of the wind, the temperature, air humidity, the atmospheric pressure, visibility, and precipitation. This change in local weather is generally associated with the movement and evolution of systems of a certain size, such as depressions and anticyclones, fronts and hurricanes. Meteorological parameters are collected daily in meteorological conditions. This observation of the fact that the atmospheric composition and other characteristics of the planet's surface have been progressively disturbed by human activity has caused worries about the future of global climate, constituting one of the biggest problems today. Concentrations of pollutants in ambient air depend on the emissions of pollutants, which are also influenced by atmospheric phenomena, which play an important role in the processes of transport, transformation, and dispersion of pollutants into the atmosphere. These processes are affected by local topography and by meteorological factors such as atmospheric pressure, temperature, solar radiation, precipitation, and wind. Practically, all the energy intervening in the atmospheric processes comes from the sun and is transferred in the form of electromagnetic waves. A strong solar radiation, associated with high temperatures, contributes to the formation of photochemical pollutants such as ozone (O_3) [3]. Temperature intervenes in the chemical component of pollutants and plays an important role in the vertical dispersion in the atmosphere. During summer, high temperatures promote O_3 formation, and in winter, temperature differences between day and night can cause thermal inversions and pollution peaks [4]. The atmospheric stability determines the local convective processes being characterised by the vertical temperature gradient that limits the vertical mixture of pollutants in case of a thermal inversion. The air temperature tends to decrease in altitude; however, under certain conditions, a thermal inversion may occur, that is, an increase in temperature, creating a layer of hot air that prevents polluted air near the ground from rising and disperse [5]. The general air flow over the planet is induced by large-scale atmospheric pressure variations (macrometeorology). These pressure variations essentially result from the differential heating of the atmosphere. The intensity of these pressure systems, their normal location or their trajectories determine the distribution of winds in a given area. The wind is a meteorological factor with direct and determinant effects on the dispersion conditions of the pollutants. The wind velocity determines the mechanical turbulence production, which is responsible for the local dispersion. The lack of wind favours the concentration of pollutants, and moderate wind conditions favour its dispersion; however, the strong wind can cause the appearance of pollution located in the direction of the prevailing winds. The low-pressure situations generally correspond to a strong turbulence in the atmosphere which favours the dispersion of the pollutants. In high pressure conditions (anticyclone), characterised by low wind, air stability does not allow the dispersion of pollutants, causing pollution to be concentrated near the ground. One way both gases and particles are removed is by dry deposition. This process involves two steps: (1) the downward movement of the molecules or particles of the pollutants until their collision with the elements of the surface; (2) its absorption or adsorption on these elements. In order for dry deposition to occur, there is a need for a downward mass flow of the pollutant to be deposited, the intensity of which depends on its atmospheric concentrations. On the other hand, the efficiency of the vertical transfer process depends on the intensity of the atmospheric turbulence in the layer under consideration, and the efficiency of the adsorption/adsorption depends on the deposited compound and the nature of the surface on which it is located made. Although both the gases and the particles undergo dry deposition, qualitative differences between their deposition processes are expected. In the case of particles of less than 1 μm, it is found that they behave essentially as gases in relation to the diffusion process. In the case of larger particles, its motion is independent of the motion of the air molecules, with each particle reaching a terminal velocity, which increases greatly with its size.

4. Particulate matter

4.1. General characteristics

Particles may be characterised as being a complex set of substances, minerals or organic substances, which are suspended in the atmosphere in liquid or solid form. Its size can range from a few tens of nanometres to a hundred micrometres. Particles are emitted into the atmosphere from a wide range of anthropogenic sources, the most important being the burning of fossil fuels, road traffic, and certain industrial processes. PM may also be emitted from natural sources such as volcanoes, forest fires or are the result of wind erosion on the soil and water surfaces. In urban areas, road transport is considered to be the largest source of PM, with the highest concentration along roadways. These substances are not only the result of direct emissions from vehicle exhaust, but also from tire wear and braking and dust resuspension. In general, diesel vehicles emit a larger amount of fine particles per vehicle than petrol vehicles [5]. The composition of airborne particles is very variable, reflecting the wide variety of emitting sources and the fact that they are continuously altering as a result of their interaction with other constituents of the atmosphere. The coarse fraction contains abundant elements of the earth's crust and marine salts, such as alum, calcium, iron, potassium, and silica, while the fine fraction is mainly composed of sulphates, nitrates or ammonia, carbon, organic compounds, and metals, mainly from the burning of fossil fuels and numerous industrial processes [6]. It is known that the smaller the particles, the greater the likelihood of penetrating deeply into the respiratory tract and the greater the risk of inducing negative effects. The finer particles being smaller than 2.5 μm ($PM_{2.5}$) reach the pulmonary alveoli and interfere with gas exchange. Chronic exposure to particles contributes to the risk of development of respiratory and cardiovascular diseases, as well as lung cancer. Suspended particulates are also an effective transport vehicle for other atmospheric pollutants that attach to their surface, especially hydrocarbons and heavy metals. These substances are often transported to the lungs where they can then be absorbed into the blood and tissues. The effects of soiling on buildings and monuments are the most obvious effects of particulate matter on the environment [5]. Atmospheric particles are associated with various health problems ranging from pulmonary to cardiovascular problems and may even lead to death. One of the effects of prolonged exposures to high particle levels is a significant reduction in the expected life expectancy. The most serious effects are usually among the most vulnerable groups, such as children, the elderly, and asthmatics [3]. The term particle generally refers to a diverse and complex set of organic and inorganic substances. Particles are a considerably large group of airborne pollutants, which may be in the liquid or solid state and originate from distinct sources such as automobiles, steelworks, thermal systems, heating systems, cement plants, volcanoes, deserts, and oceans. In general terms, the term particle may be taken to be any three-dimensional discontinuity in the liquid or gaseous phase in a dispersed system. However, in terms of air pollution, a particle can be defined as a solid, liquid or solid liquid dispersed matter, and the individual aggregates having a diameter between 0.0002 and 500 μm [7]. Thin particles are considered if their aerodynamic diameter is less than 1.0 μm and are mainly emitted from sources of anthropogenic origin as combustion processes, including exhaust emissions of vehicles. The particle classification can be made based on two criteria: the shape mechanism or the physical dimension. In the case of classification according to the mechanism of forming the particulars can be classified as primary particles, which are emitted directly, and secondary particles or particles that are those formed from precursor gases existing in the atmosphere, through a mechanism of form-particle. Both the so-called primary and secondary particles are subject to growth and transformation mechanisms, since secondary material may also be formed on the core of the existing particle. In the case of classification according to the physical dimension, which is the most used classification, it ranges from a few nanometers (nm) to tens of micrometers (μm) in diameter. Dimension is a very important feature and has implications for form, physical and chemical properties, transformation, transport, and particle removal, from the atmosphere. Knowing that the particles in suspension in the atmosphere vary considerably in size, composition, and origin, it is important to classify the particles for their aerodynamic properties, since these properties, besides being responsible for the transport and removal of particles from the air, also generate their deposition in the respiratory system, being also associated to the chemical composition and origin of the particles [8]. Thus, particle size is usually characterised by its aerodynamic diameter, which refers to the diameter of a sphere of uniform density and with the same settling velocity of the particle in question [9].

4.2. Classification of the particles

In urban environments, mass and particle composition tend to be confined to two major groups: coarse particles (larger particles) and fine particles (smaller particles). The boundary between these two classes of particles is generally between 10 and 1 μm. However, this limit between coarse and fine particles is generally fixed, by convention, to 2.5 μm of aerodynamic diameter. The standard PM_X (USEPA terminology) refers to particles with a diameter less than x μm, whereby $PM_{2.5}$ refers to particles with a diameter of up to 2.5 μm. Smaller particles (fine particles) include secondary aerosols, formed from gases in the atmosphere through the gas-particle formation mechanism (gas-particle conversion), and also contain particles that result from combustion processes and organic recondensed vapours and metallic. Due to their small size, they are easily inhaled, depositing in the lower respiratory tract and causing numerous, essentially respiratory, health problems. The term 'Total Suspended Particles' (TSP) refers to a mass concentration of particles of less than 50 μm in diameter, and the term 'ultrafine particles' refers to particles of diameter less than 100 nm (0.1 μm). Larger particles (coarse particles) usually contain materials from the earth's crust and dust from roads and industry. In urban environments, the largest number of particles is found in very small sizes, less than 100 nm. However, these ultrafine particles (UFP) often contribute a small percentage to the total mass of the sample, contributing more than 90% of the number of particles. Particles with a diameter of less than 1 μm (PM_1) have a size that allows them to penetrate deeply into the respiratory and circulatory system carrying toxic elements and compounds [3]. Other thermal usually used in black carbon (BC) with a primary aerosol emitted directly at the source from incomplete combustion processes (fossil fuel and biomass burning) and so a several part of atmospheric BC is of anthropogenic origin. Chemically, BC consists of pure carbon in several linked forms.

4.3. Mechanisms of particle formation

Particles with a diameter greater than 2.5 μm (coarse particles) are produced mechanically by the breaking of larger solid particles, which may include dust originating from agricultural processes, glues carried by the wind from the exposed soil, dirt roads, or dust from other processes such as mining or quarrying. In turn, also the road traffic produces dust and turbulence that causes rise and can shake the road dust. Also at coast-to-coast locations, evaporation of the sea water can produce particles of this size. Also, pollen grains, mould and plant spores, and insect parts are included in this larger size range. The amount of energy needed to break down the elements referred to in smaller particles increases as the size decreases. This results in a lower limit for the production of these coarse particles of approximately 1 μm. Minor particles (fine particles) are formed, for the most part, from gases, through two distinct processes according to its size, nucleation and condensation. In nucleation, the smallest particles, less than 0.1 μm, are formed by the condensation of substances formed by vaporisation at high temperature or by chemical reactions in the atmosphere. Particles in this range grow by coagulation, that is, the combination of two or more particles to form a larger particle, or by condensation, that is, condensation of gas or vapour of molecules on the surface of existing particles. Coagulation is more efficient for large particle numbers, and the condensation is more efficient for large surface areas. Therefore, the efficiency of both coagulation and condensation mechanisms decreases with increasing particle size, which effectively produces an upper limit such that the particles do not grow by these processes to more than about 1 μm. Thus, this type of particles tends to 'accumulate' between 0.1 and 1 μm. In condensation, particles below 1 μm can be formed by condensation of metals or organic compounds, which are evaporated in combustion processes, or can also be produced by gas condensation arising from atmospheric areas. For example, sulphur dioxide is oxidised into the atmosphere to form sulphuric acid (H_2SO_4), which can be neutralised by ammonia (NH_3) to form ammonium sulphate. Nitrogen dioxide (NO_2) is oxidised to nitric acid (HNO_3) which in turn can react with NH_3 to form ammonium nitrate (NH_4NO_3). The particles produced by these gases in the atmosphere are called secondary particles. Sulphates and nitrate particles are usually the predominant component of these fine particles [10]. Suspended particles in the environment typically have a modal type distribution with respect to size (diameter), which means that the total mass of the particles tends to be concentrated around one or more distinct points. The modal distribution character in the particle size results from the equilibrium of the particle formation processes on the one hand and on the other side of the particle removal processes from the atmosphere. Thus, this modal distribution of the diameter of the particles around one or two characteristic points varies depending on the age of the aerosol and the proximity of emission sources of particles of different types. Other important aspects in the definition of particle concentrations in the atmosphere are meteorological variables such as wind speed and direction, atmospheric temperature, precipitation, and height of the atmospheric boundary layer. Higher particle concentrations are often recorded during atmospheric weather conditions, especially in thermal inversion situations with low wind speed, and also because the physical and chemical processes of particle formation are governed largely by meteorological variables [11].

4.4. Point sources and area sources

There are several types of source emission related to PM. Particulate sources designated by point sources include various types of facilities such as power plants, industrial plants, municipal waste incineration plants, paper mills, various fossil fuel combustion plants, and domestic heating installations. These sources are often considered as point sources (thermal and industrial plant chimneys), but may also be considered as an area source, such as residential combustion plants. The physical and chemical characteristics of the particles emitted from these source categories depend on the combustion process itself and on the type of fuel burned, presenting quite different physical, chemical, and dimensional characteristics depending on the process combustion. For noncombustion emissions, the main industrial processes that may contribute to the emission of particulate matter to the atmosphere include metal processing and chemical processing plants, processing and handling of building materials or for industry. Particulate emissions originating from this type of source are often derived from fugitive emissions, which are not controlled but are instead released in an inhomogeneous form. The type of particles and their physical and chemical properties also depend on the processes by which they are emitted, and it is not at all possible to generalise their characteristics. The knowledge of the relation of the various dimensions of the particles in a certain sample of atmospheric air is important to try to characterise the origin of the emission sources of these same particles. Some recent studies [12] have devoted themselves to studying this relationship, based on samples collected from 31 locations in Europe, concluding that the diameters relationships showed similarities for all locations [13].

4.5. Remote sources

Particles in the form of dust of natural origin carried by the wind can contribute to the existence of high concentrations of larger particles (coarse particles) and fine particles. In some cases, the particles are found in locations hundreds or thousands of kilometres from their origin. It has been proven the strong contribution of the wind in transporting dust from the desert to remote sites of its origin. For example, in Southern Mediterranean countries, such as Portugal, there are frequent 2–4-day transient episodes of transporting dust from the Sahara desert every year, resulting in levels exceeding 25 $\mu g/m^3$ expected daily concentrations of PM_{10} and $PM_{2.5}$ [14].

Resuspension of particles is the term given to the re-entry, in atmospheric air, of particles previously deposited and their re-entrainment into the atmosphere. It is a complex process that can be triggered by mechanical disturbances, such as wind, traffic-induced turbulence, tire stress, and construction activities. The so-called 'road dust' is an agglomeration of particles originating from various anthropogenic and biogenic sources. On the roads, this dust of diverse origins accumulates on the roadsides, near the sidewalk and along the central divisions. Resuspension, deposition, entrainment on and off the road, and emission of new particulates are a dynamic particle emission 'source' and 'well' mechanism that characterises road traffic. Roads are one of the largest source emitting particles in urban environments. Several studies have also shown that resuspension of this element is the predominant source of larger particles (coarse particles) in locations of intense road traffic causing the impact of resuspension on the concentration of particles in the atmospheric air is of great importance. Road dust may also act as a repository for the various elements of anthropogenic particulate sources, and resuspension may function in certain locations as a re-emission, thereby contributing to the increase in the atmospheric concentration of these elements. The plausibility of this theory is supported by studies where it has been shown that larger particles are more easily resuspended by wind and road traffic, and that deposited materials are more susceptible to resuspend, if associated with larger host particles. Fine particles can remain in suspension for much longer than coarse particles and this may result in a greater spatial impact on atmospheric concentrations of particles, and secondly the fine fraction of resuspended particles is more likely to contain constituents of anthropogenic origin, potentially more toxic, than the fine particles of natural origin [15].

4.6. Contribution of road traffic to PM

In urban environments, 90% of the concentration of atmospheric particles emitted by road traffic corresponds to the fraction of particles smaller than 1 mm (0.001 μm). This fact can affect human health, so it is important to study PM_1 in high traffic areas [16]. Particulate emissions from road traffic are the result of a large number of processes, such as the combustion products of gasoline, diesel, and gas engines, products originating from vehicle oil, tire rubbers, braking system, bearings, car body, road material, and dust release from road and ground [17]. Traffic is in fact an important source of both smaller (fine particles) and larger (coarse particles) particles, but is also a source of condensable organic gas emissions and an important source of nitrogen (NO_X), which subsequently form nitrate (secondary) aerosols. Particles of condensed carbonaceous material are emitted mainly by diesel vehicles, but also by gasoline vehicles with run-down performance [13]. Particulates originating from diesel engines are mainly carbonaceous agglomerates less than 100 nm in diameter, while the particles emitted by gasoline vehicles are mainly smaller carbonaceous agglomerates, ranging from 10 to 80 nm [18]. Although it is not possible to generalise conclusions about the association of the various elements present in atmospheric particles, with their origin in road traffic, some elements have been frequently associated with them. These elements include copper (Cu), zinc (Zn), lead (Pb), bromine (Br), iron (Fe), calcium (Ca), and barium (Ba) [18–20]. However, the emissions of many of the metal elements originating from road traffic are not due to the exhaust emissions but to other sources of the vehicle such as tires, brakes, and other parts of the vehicle [17, 21, 22]. Studies have concluded that $PM_{2.5}$ containing sulphur in their composition is mainly from the combustion of coal and exhaust emissions from vehicles [22]. Polycyclic aromatic hydrocarbons (PAHs) are organic compounds, formed by at least two fused aromatic rings, entirely made up of carbon and hydrogen [23]. They can be found in many urban air components and are a health concern, mainly because of their carcinogenic and mutagenic properties. A negative correlation of PM_1 with the wind speed was obtained due to the wind dispersing the particulate matter from the atmosphere. Regarding relative humidity, it was found to have a positive correlation with PM_1, which can be attributed to the influence of free and clean masses of troposphere air. For the ambient temperature and solar radiation, a negative correlation was calculated, perhaps associated with stagnation and cold fronts. A positive correlation of PAHs with relative humidity and a negative correlation with solar radiation, ambient temperature, O_3 and NO shows that PAHs degrade through photolysis and chemical reactions with these pollutants. Analyses of the composition of PM_1 with PAHs indicated that these had their origin especially in diesel and gasoline emissions, as well as the combustion of wood, lubricating oils, and fossil fuels [16].

4.7. Consequences of PM in atmosphere

Particulate matter (PM) is one of the most relevant air pollutants globally. In humans, adverse effects associated with many cases of exposure to high concentrations of aerosols (mortality, morbidity, respiratory, and cardiovascular problems) are well established. However, the mechanisms involved are still not well known [24]. In recent years, several scientific studies have attempted to correlate the outcome of possible adverse health effects due to the exposure to PM levels in atmospheric air [25]. Numerous epidemiological and toxicological studies have recently been developed to try to understand what kind of particles and which dimensions lead to the most detrimental effects on human health. In chemical terms, some studies indicate that the toxicity of the particles is mainly due to the organic compounds around the particle, and other studies point to the coal core of the particle as the main factor of toxicity. In terms of size, a considerable number of authors correlate the health effects with the mass concentration of particles, and other authors point out the importance of the concentration of ultrafine particles in atmospheric air in the negative consequences for health. Several epidemiological studies have shown the strong correlation between morbidity (or morbidity) and mortality with the concentration of fine particles in urban environment. These refer to particles as the air pollutant with the most detrimental consequences to health, followed by ozone (O_3). Some studies indicate that even concentrations below the current recommended air quality levels may pose a health risk. It is believed that the effects of fine particles on health are caused after their inhalation and penetration into the lungs. Several studies indicate that both chemical and physical interactions with lung tissues can cause irritation or damage to the lungs. The smaller the size of the particles, the more they can penetrate the lungs. Annual mortality levels are associated with the concentration levels of $PM_{2.5}$, which in Europe represent 40–80% of the mass concentration of PM_{10} in ambient air. However, the larger particle fraction (from 2.5 to 10 μm) of PM_{10} also has negative impacts on human health and affect mortality, although increasing evidence points to $PM_{2.5}$ having an adverse impact [11]. Continued exposure to atmospheric particles contributes to the risk of developing cardiovascular and respiratory diseases as well as lung cancer. The mortality associated with air pollution is about 15–20% higher in cities with high levels of pollution compared to relatively less polluted cities. From this air pollution, numerous studies show that the most critical pollutant is the particles. For example, in the European Union (EU) studies indicate that the average life expectancy is 8.6 months lower due to exposure to $PM_{2.5}$ resulting from human activities. Particle pollution (especially fine particles) contains microscopic solids or liquids, which, being too small, can penetrate deeply into the lungs and cause serious health problems. Numerous scientific studies have correlated exposure to particulate matter with a number of health problems, including: increased respiratory problems such as irritation of the airways, coughing or difficulty breathing; decreased pulmonary function; worsening of asthma cases; development of chronic bronchitis and irregular heartbeat; non-fatal heart attacks and premature death in people with heart or lung disease. People with heart or lung disease, children, and the elderly are the groups most likely to be affected by exposure to particulate pollution. As mentioned above, two types of studies have been developed with regard to the health consequences of exposure to air pollutants, so this also applies to particulate matter, epidemiological studies, and toxicological studies. Epidemiological studies are global studies that seek to study the cause-effect relationship of a given disease, most often using statistical analysis tools. In the case of particles, these studies seek to analyse the consequences of exposure of a given population to a particular concentration level or to a particular type of particle. Within this type of studies, two degrees of incidence have been analysed: morbidity, which can be defined as the rate of carriers of a given disease in relation to the total population studied, and mortality. Alternatively, toxicological studies are studies that attempt to analyse the harmful or adverse effects that a certain toxic agent (chemical) has on the organism. There are two types of toxicological studies, experimental toxicology, which uses animals to try to understand the mechanism of action and consequences for the body, and analytical toxicology that aims to identify/quantify toxic elements in organs such as liver, kidneys or matrices such as blood, urine, or saliva [26]. Combustion of biomass by the domestic sector (burning of fuels such as wood and coal) and emissions from road vehicles in urban centres are the sectors that are the main direct sources of particulate emissions. Agriculture is a sector with high contribution to ammonia emissions, which is one of the pollutants that contributes most to the formation of secondary particles. Particles can also affect the climate, promoting heating or cooling of the planet, depending on its chemical composition. One such case is soot containing black carbon, found mainly in fine particles, resulting from the incomplete burning of fossil fuels and biomass, contributes to changes in the climate because it absorbs the energy of the sun promoting the heating of the atmosphere [27].

5. Monitoring networks

National or regional air quality authorities have in their area of jurisdiction a set of fixed sampling points for the continuous measurement of PM concentration. For their realisation, they have stations equipped with automatic analysers of measurement PM concentration. The main objective in the process of selecting the localization of air quality stations is to obtain information with the greatest possible representation of the surrounding area, since the location of the air quality measurement stations can directly affect the conclusions from the analysis of the results. Measuring stations that may be influenced by very particular characteristics of the locations where they are installed may no longer be representative and require the analysis of complementary information from other stations or other assessment methods. This reason makes it necessary for information users to have systematised data on the particular conditions of installation and location of each station. The selection of monitoring sites should take into account the diversity of techniques and materials used in the construction of buildings, as they differ from country to country and in larger countries, differ within the same country, in order to have representative sites in each area. In urban areas, the deterioration of materials is affected by the levels of pollution observed. Three locations can be chosen to represent: the highest level of urban pollution (usually near the centre of the city), an average background level, and a traffic hot spot. In industrial zones, two sites may be selected to represent a medium level and a higher level of pollution. The air quality stations are classified taking into account the area where they are located (surrounding environment) and the type of emission source that influences the air quality levels measured in that location and may be in relation to the type of zone or type of environment encompassing three possible classes: urban, suburban, and rural and as to the type of dominant emission source containing three possible classes: traffic, industrial, and fund. The classification of stations in the various typologies is relevant for analysing the air quality data, as it makes the data comparable at local or global level [28]. Each country develops its plan for monitoring air quality and deposition in order to be representative of ecosystem exposure to air pollution. Since concentrations of air pollution and deposition also vary greatly, a monitoring program is required that includes a large selection of ecosystems, that is, regions and areas within each region [29].

http://dx.doi.org/10.5772/intechopen.79695

6. Methods of PM measurement

For the continuous monitoring of PM in the air, sophisticated and very sensitive equipment is essential. For the quality control of the information generated, periodic calibrations, corrective and preventive maintenance, and evaluation of the representativeness and validity of the data obtained through the statistical analysis and monitoring of the historical trend of the pollutant at the location in question and of the analyser behaviour are necessary. The estimation of the uncertainties that must be evaluated in the calibration of the equipment for the monitoring of environmental data presents a level of difficulty because the concentration range of pollutants found in an environment is very close to the detection limits of the equipment available in the market and due to the number of factors that interfere with the measurement of the pollutant in question. The assurance of the presented results is based on a good detection of errors and inconsistencies occurred in the procedures and analytical or sampling methods. Failure to do so may lead to misinterpretations or misconceptions [30]. The most common method of PM measurement and monitoring is done by a beta particle analyser by absorption of beta radiation and sampler. This method applies to the automatic measurement of continuous particulate matter, based on the absorption of the β radiation emitted by a radioactive source by the particles deposited in a filter. The particles are deposited on a fibreglass tape, which is traversed by constant flows of ambient air for pre-programmed time periods or cycles. This tape is located between the radioactive source and the Geiger-Muller radiation detector. The uniform distribution of the deposited particles on the surface of the filter allows to obtain the relation between the total mass deposited and the number of counts registered by the detector. Knowing the number of counts recorded with the filter before the aspiration of the ambient air and the number of counts recorded with the filter after the deposition of particles, it is possible to determine its mass in $\mu g/m^3$.

7. Strategic aspects for decision making

Information on atmospheric pollution levels in general and external air quality (AQ) in particular is a widely discussed and developed aspect both in the scientific community and in the general population. However, an effort has been made by the scientific community to develop models that somehow predict and simulate PM concentration in the environment. An even more recent and ambitious step refers to the modelling of the human exposure to this pollutant, relating the activity of the individual and his time of permanence in this space with the environments in which he is in terms of PM concentration [31]. It is therefore essential that mitigation and mitigation measures should be implemented through National and International Plans and Programs that will lead to significant benefits for atmospheric PM levels to be significantly reduced with benefits to health [32]. These measures include:

- Intervening on the effectiveness of air quality and emission legislation by strengthening their implementation;
- Overall and global reduction of PM emissions from industrial and domestic sources;

- Modernization of monitoring and data transmission and control of PM;
- Creation of significant benefits in the acquisition of new vehicles that are less polluting or even non-polluting and more energy efficient;
- Promotion of access to shore-based electricity (preferably from renewable energy sources) by berthed ships;
- Selection of forest species and practices less vulnerable to storms and fires, greater penalties for the fire set;
- Reduction in the use of nitrogen (N) in agriculture due to the excessive use of nitrogen fertilisers and nitrogen content in animal feed;
- Implement improved monetary assessment of impacts on ecosystems and analyse the costs and effectiveness of measures implemented to improve QA;
- Rationalise and optimise the global transport system to reduce CO_2 emissions.

8. Conclusions

Improving global air quality and reducing the atmospheric concentration of PM is essential to achieve greater control of air pollution at source and to minimise its effects in order to preserve human health, the environment, and materials. Legislative measures and other instruments (including Plans and Programs) have been adopted at National and Community levels for several decades. For these measures and instruments to be implemented and enforced, it is necessary to implement the updated air quality policy, making use of the latest scientific knowledge and appropriate management of commitments and synergies between climate and environmental objectives in order to that it cannot only enhance ecological and climatic resilience, but also achieve important socio-economic benefits, including public health.

The control at source of emissions of air pollutants in general and PM in particular can be done through the installation and use of best available non-pollutant techniques and equipment. Also, measures to prevent air pollution include reducing emissions, as switching to cleaner fuels, alternative use of renewable energy sources, and more efficient and cleaner new energy sources (wind, water, solar, and biomass valorisation of agriculture) in response to considerations of the climate or security of supply, the use of less polluting means of transport, leading in particular to a substantial decrease in PM_{10} and dangerous emissions. In this chapter, a general overview of variables and key factors that identify, relate, and understand the diverse and multidisciplinary variables that contribute to PM concentration in urban environments associated with health impacts were identified and described. The main multidisciplinary aspects, namely meteorology, urban geometry, road traffic, industries, air concentration measurements, and health were also presented and discussed. The main strategic aspects for decision making related to airborne PM impact on health were also discussed.

Author details

João Nuno Pinto Miranda Garcia[1,2]*, Joana Borrega[1] and Luís M. R. Coelho[1,2]

*Address all correspondence to: joao.garcia@estsetubal.ips.pt

1 Polytechnic Institute of Setubal (IPS), EST, Setubal, Portugal

2 CINEA-IPS, Centro de Investigação em Energia e Ambiente, Setúbal, Portugal

References

[1] WHO. Update of WHO Air quality guidelines 2008. WHO Regional Office for Europe; 2008

[2] Hackley VA, Ferraris CF. The Use of Nomenclature in Dispersion Science and Technology, NIST Recommended Practice Guide. NIST. Washington, United States of America: Special Publication; 2001. pp. 960-963

[3] HEI. State of Global Air 2018. Special Report. Boston, MA: Health Effects Institute; 2018

[4] Respostas para artigo de Saúde Ambiental Jornal Água & Ambiente [Internet]. 2015. Available from: https://www.apambiente.pt/_cms/view/page_doc.php?id=486 [Acessed: October 12, 2016]

[5] O Ar e os Poluentes Atmosféricos [Internet]. 2014. Available from: http://www.ccdr-lvt.pt/pt/o-ar-e-os-poluentes-atmosfericos/8082.htm [Accessed: October 12, 2016]

[6] Nogueira L, Mesquita S. Relatório de avaliação da qualidade do ar ambiente na região de Lisboa e Vale do Tejo no período 2001-2013; 2014

[7] Iijima, Tago H, Kumagai K, Kato M, Kozawa K, Satob K, Furuta N. Regional and seasonal characteristics of emission sources of fine airborne particulate matter collected in the centre and suburbs of Tokyo, Japan as determined by multielement analysis and source receptor models. Journal of Environmental Monitoring. 2008;**10**:1025-1032

[8] Garcia J, Teodoro F, Cerdeira R, Coelho LMR, Kumar P, Carvalho MG. Developing a methodology to predict PM_{10} concentrations in urban areas using generalized linear models. Environmental Technology. 2016. pp. 1-34. DOI: 10.1080/09593330.2016.1149228

[9] Kumar P, Morawska L, Birmili W, Paasonen P, Hu M, Kulmala M, Harrison RM, Norford L, Britter R. Ultrafine particles in cities. Environment International. 2014;**66**:1-10

[10] Cordeiro GM, Demétrio CGB. Modelos Lineares Generalizados; 2007

[11] Pohjola M, Kousa A, Aarnio P, et al. Meteorological interpretation of measured urban $PM_{2.5}$ and PM_{10} concentrations in the Helsinki metropolitan area. In: Longhurst JWS, Brebbia

CA, Power H, editors. Air pollution VIII. Wessex: Institute of Technology Press; 2000. pp. 689-698

[12] Van Dingenen R, Raes F, Putaud J, Baltensperger U, Charron A, Facchini M, Decesari S, Fuzzi S, Gehrig R, Hansson H. A European aerosol phenomenology 1: Physical characteristics of particulate matter at kerbside, urban, rural and background sites in Europe. Atmospheric Environment. 2004;**38**:2561-2577

[13] Vallius M. Characteristics and Sources of Fine Particulate Matter in Urban Air. Health. Kuopio, Finland: National Public Health Institute, Department of Environmental Health; 2005

[14] Rodriguez S, Querol X, Alastuey A, Viana M, Alarcon M, Mantilla E, Ruiz CR. Comparative PM_{10}-$PM_{2.5}$ source contribution study at rural, urban and industrial sites during PM episodes in Eastern Spain. The Science of the Total Environment. 2004;**328**:95-113

[15] Vouitsis E, Ntziachristos L, Pistikopoulos P, Samaras Z, Chrysikou L, Samara C, Papadimitriou C, Samaras P, Sakellaropoulos G. An investigation on the physical, chemical and ecotoxicological characteristics of particulate matter emitted from light-duty vehicles. Environmental Pollution. 2009;**157**:2320-2327

[16] Agudelo-Castañeda D, Teixeira E. Seasonal changes, Identification and Source Apportionment of PAH in $PM_{1.0}$; Atmospheric Environment. Vol. 96. 2014. pp. 186-200

[17] Laschober C, Limbeck A, Rendl J, Puxbaum H. Particulate emissions from on-road vehicles in the Kaisermuhlen-tunnel (Vienna, Austria). Atmospheric Environment. 2004;**38**: 2187-2195

[18] Morawska L, Zhang J. Combustion sources of particles. 1. Health relevance and source signatures. Chemosphere. 2002;**49**:1045-1058

[19] Sternbeck J, Sjödin A, Andreasson K. Metal emissions from road traffic and the influence of resuspension—Results from two tunnel studies. Atmospheric Environment. 2002; **36**:4735-4744

[20] Kemp K. Trends and sources for heavy metals in urban atmosphere. Nuclear Instruments and Methods in Physics Research Section B: Beam Interactions with Materials and Atoms. 2002;**189**:227-232

[21] Lough GC, Schauer JJ, Park JS, Shafer MM, Deminter JT, Weinstein JP. Emissions of metals associated with motor vehicle roadways. Environmental Science & Technology. 2005;**39**:826-836

[22] Adachi K, Tainosho Y. Characterization of heavy metal particles embedded in tire dust. Environmental International. 2004;**30**:1009-1017

[23] Decree-Law no. 102/2010, September 23

[24] Pateraki St, Asimakopoulos DN, Flocas HA, Maggos Th, Vasilakos Ch. The role of meteorology on different sized aerosol fractions (PM_{10}, $PM_{2.5}$, $PM_{2.5-10}$). Science of the Total Environment. Vol. 419. January 2012

[25] Garcia J, Cerdeira R, Coelho L, Kumar P, Carvalho MG. Important variables when studying the influence of particulate matter on health. In: Chapter of Book Particulate Matter: Sources, Impact and Health Effects. Nova Publishers; 2017

[26] Moreira S. Procedimentos Regionais de Informação e Alerta no Âmbito da Qualidade do Ar, Relatório de Ambiente e Saúde; 2010

[27] Ferreira F, Seixas J, Barroso J, et al. Estratégia Nacional para o Ar 2020 (ENAR 2020): Emissões Atmosféricas e Qualidade do Ar Ambiente: Enquadramento e Diagnóstico; 2015

[28] Fernandes A, Guerra M, Ribeiro R, et al. Relatório do Estado do Ambiente 2015 (REA 2015). Lisbon, Portugal: Agência Portuguesa do Ambiente; December 2015

[29] Matos J, Marques A, Brantes J, et al. Manual de Métodos e de Procedimentos Operativos das Redes de Monitorização da Qualidade do Ar – Amostragem e análise. Agência Portuguesa do Ambiente; 2010

[30] Santana G, Guerra I, Leite C, et al. Monitoramento da qualidade do ar–Estimativa da incerteza de medição de analisadores de NOX; 2011

[31] Garcia J, Cerdeira R, Coelho L, Kumar P, Carvalho MG. Influence of pedestrian trajectories on school children exposure to PM10. Journal of Nanomaterials. Hindawi Pub. Corp. Article ID: 505649. 2014:**2014**;9. DOI: 10.1155/2014/505649

[32] Lee KK, Miller MR, Shah ASV. Air pollution and stroke. Journal of Stroke. 2018;**20**(1):2-11. DOI: 10.5853/jos.2017.02894

Chapter 3

Commuting on Public Transport: Health Risks and Responses

Li Ping Wong, Haridah Alias,
Nasrin Aghamohammadi, Nik Meriam Nik Sulaiman,
Hualiang Lin and Jinshun Zhao

Additional information is available at the end of the chapter

http://dx.doi.org/10.5772/intechopen.79694

Abstract

Rapid urbanization places a huge demand for infrastructure on busy city roads, exposing commuters to the health risks of atmospheric air pollutants. Traffic-related air pollution causes significant health burdens worldwide and enormous economic losses. Both short- and long-term exposures to atmospheric air pollutants cause a range of adverse health effects on people who commute. The short-term symptoms of exposure include coughing, shortness of breath, chest pain, and headaches. Long-term exposure is associated with cardiovascular, respiratory, and lung cancer mortality. Exposure to traffic-related air pollution also has detrimental effects on psychological and mental health. Although the use of proper respiratory protection may greatly reduce inhalation of microparticulate matter commonly found at high concentrations on busy roads, its use is not common in some communities. The adverse health risks associated with traffic-related air pollution can also be reduced through diet and lifestyle modifications, and these should be encouraged. A lack of environmental health literacy may result in the underutilization of preventive resources; therefore, fostering proper delivery of information may improve the health of commuters. Attention is drawn to the need for research tailored to individual societies or countries due to the influence of innumerable factors such as culture, religion, and climate, as well as policy and governance that contribute to diversity among health impacts and local community mitigation measures.

Keywords: health risks, mitigation responses, traffic-related pollution, commuters

1. Introduction

In recent decades, Asia has experienced rapid development of industrialization and urbanization, which has resulted in accelerated growth of many suburban cities surrounding large metropolitan areas. These suburbs are home to many who work in metropolitan areas, causing a tremendous number of suburban residents each day to commute to the city for work. As a consequence, urban areas are experiencing increasing automobile use, resulting in congestion and long hours of commuting. The increasing severity and duration of traffic congestion greatly intensify pollutant emissions and degrade air quality.

It is of tremendous concern that commuters are exposed to high concentrations of atmospheric pollutants, originating from both vehicular traffic and other urban, industrial, or environmental factors, during their transit. They are exposed during their daily commute while in vehicles, waiting for transportation and walking. Most commuters spend considerable time at bus stops and stations or walk on busy roads. It was reported commuters who travel by public transport may be exposed to up to eight time as much air pollution as those who drive to work [1].

1.1. Health impacts of air pollution

Most megacities in Asia are facing acute problems due to an increase in the ambient particulate matter (PM) and nitrogen dioxide (NO_2) concentrations as a result of rapid urbanization. In Shanghai, New Delhi, Mumbai, Guangzhou, Chongqing, Calcutta, Beijing and Bangkok, the ambient PM and NO_2 concentrations were reported to frequently violate World Health Organization (WHO) guidelines [2]. Worldwide, in most of the megacities, motorized road transport is categorized as one of the largest pollution sources. Motorized transport is responsible for 70% of environmental pollution and 40% of greenhouse gas emissions in European cities [3]. In the UK, automobile pollution sources frequently violate the national ambient air quality standards [4]. Likewise, on the Asian subcontinent, some rapidly developed countries, such as Singapore, Japan, and Hong Kong, are facing equally critical street-level air pollution problems due to an increase in the number of motorized vehicles [4].

Road vehicle emissions are one of the most important sources of human exposure to air pollution. Air pollution concentrations on roads are relatively high, and commuters face unavoidable exposure during commuting as they are near the source of emission. The deleterious effects of traffic-related atmospheric air pollution on health have been documented in many studies worldwide. Emissions from road traffic result in a complex mixture of harmful air pollutants. In many areas, vehicle emissions have become the dominant source of air pollutants, including carbon monoxide (CO), carbon dioxide (CO_2), volatile organic compounds (VOCs), hydrocarbons (HCs), nitrogen oxides (NOx), and particulate matter of aerodynamic diameter 2.5 μm ($PM_{2.5}$) [5]. These toxic particles easily penetrate human airways. Inhaled $PM_{2.5}$ can reach the lung alveoli and induce local and systemic responses in the body, impacting cardiovascular and respiratory function [6]. Thus, traffic-related air pollution has been implicated in a range of illnesses related to respiratory diseases and cardiovascular complications. From a public health perspective, the health effects of air pollution are both chronic (long term) and acute (short term). Breathing in high quantities of exhaust fumes can cause short-term irritation to the respiratory tract within a few minutes of exposure. Short-term inhalation of air pollutants may exacerbate ongoing irritation such as cough, mucous buildup, and inflamed airways. Although the acute, short-term effect is of the least concern to the general public, over time, long-term exposure eventually merges with the chronic effect. Of concern, prolonged exposure over many years places tremendous stress on the body and can be detrimental to human health.

Air pollution is now the world's largest environmental health risk [7]. In 2015, the WHO released a report stating that at least one in eight deaths worldwide is caused by air pollution. On a global scale, air pollution accounts for an estimated 9% of deaths due to lung cancer, 17% due to chronic obstructive pulmonary disease, more than 30% due to ischemic heart disease and stroke, and 9% due to respiratory infections [8]. It is estimated that approximately 80% of the world population lives in environments with pollution levels exceeding the air quality guideline (AQG) established by the WHO [9]. It is also a well-established fact that air pollution shortens life expectancy. The high level of particles in air pollution is related to lung cancer risk, cardiovascular disease, and mortality. In many countries in Asia, concentrations of ambient air pollutants exceed levels associated with increased risk of acute and chronic health problems. In particular, studies have shown that exposure to air pollution in traffic has been associated with chronic health effects, particularly cardiovascular and respiratory diseases [10–12]. Much evidence has also been found showing the high prevalence of respiratory disease symptoms and asthma exacerbation among those who reside near high-traffic roads [13, 14].

On the Asian continent, China, as one of the fastest developing countries, is battling the health impact associated with air pollution. Annual average $PM_{2.5}$ concentrations in Chinese megacities exceed the WHO's guideline of 10 $\mu g/m^3$, and the corresponding black carbon (BC) concentrations are approximately 5 $\mu g/m^3$ [15]. The primary sources of air pollution in China are industrial output, coal and biomass combustion, and traffic. Nevertheless, a recent report noted that emissions from heavy urban traffic are the main contributors to urban air pollution in China [16]. The 2010 Global Burden of Disease reported that exposure to air pollutants is the fourth leading health risk factor for Chinese people [17]. Likewise, air pollution in China is also associated with elevated rates of mortality whereby an estimated 350,000 to 500,000 premature deaths were reported to be linked to air pollution [18]. Similarly, in Jakarta, one of the most polluted cities in the world, air pollution was largely associated with motor vehicles, particularly emissions largely created by diesel vehicles [19].

1.2. Economic cost of health impacts of air pollution

Most countries globally and also those in Asia suffer enormous economic loss related to health effects of air pollution on the public. For instance, in China, the economic costs of the health impact of air pollution exceed expectations. It has been estimated that in 2005, China's direct welfare loss from mortalities associated with ozone (O_3) and PM exposure was US$42 billion [20]. In Jakarta it was reported that the health cost of air pollution in 1999 reached US$220 million [19]. Likewise, in other countries in Asia, Singapore reported a total economic cost of

US$3662 million associated with the health impact of particulate air pollution [21]. In Malaysia, a remarkable increase in the hospital admission was noted during almost all haze episodes. In the recent 2013 haze, the cost of illness peaked at MYR410 million in Malaysia [22]. Of note, the cost of air pollution exceeds estimations. The economic impact of air pollution on human extends beyond health. It also has a tremendous impact on social and emotional well-being and daily activities of the affected community. Other important costs of the social effects of air pollution, such as loss of vegetation, and the resulting infertility of the surrounding land, loss of productivity, and decreased work efficiency have not been accounted for.

Considering the immense economic impact of air pollution, the affected communities, particularly the commuters who are regularly exposed to atmospheric air pollution, should be equipped with knowledge of the health impact of air pollution and preventive measures to reduce the health risks associated with exposure to air pollution. A concerted effort on the part of the public to practice mitigation measures is important in order to reduce the economic loss caused by the effects of air pollution.

1.3. Improving environmental health literacy

While effective policies to reduce traffic emissions at their sources are clearly preferable, the evidence supports the benefits of individual personal actions to reduce exposure and health risks associated with traffic emissions. It is well established that knowledge and positive attitudes enhance the population's interest in adopting healthy preventive behaviors. In the context of air pollution and the adverse health impact, environmental health literacy needs to be enhanced. Environmental health literacy is a measure of individual understanding of specific risks, which then leads to broader understanding, including strategies that empower people to reduce or eliminate environmental exposures that can harm their health [23]. Inadequate environmental health literacy, particularly regarding traffic-related air pollutants, may impair mitigation practices, thus leading to health impairment. This is because the level of an individual's awareness and concern has demonstrable effects on whether individuals are willing to carry out self-prevention of exposure to atmospheric air pollution during the daily commute. There was substantial evidence reporting individual- and community-level behavior change in response to environmental exposure and education about exposure [23]. The former is particularly true in Ningbo, where, despite experiencing a relatively low level of exposure to ambient pollution compared with other cities, residents showed a higher awareness about ambient air pollution and its adverse impact on health, which was due to previous episodes of heavy smog they experienced [24]. This implies the importance of experience in shaping health literacy and its implications for behavioral change.

Nevertheless, a review of the literature found that, in many megacities in China and India, as well as Malaysia, despite heightened air pollution, on the whole, there are still substantial important knowledge gaps among the population regarding causes, effects on health, and prevention practices that need considerable attention [25–27]. Efforts are needed from around rapidly developing cities in Asian regions affected by air pollution to set goal-bridging gaps and advance the population's knowledge about air pollutants and health outcomes and, most importantly, on pollution prevention practices that reduce or eliminate risks.

1.4. Community-level policy and practice in reducing traffic-related air pollution

Policy makers and urban planners at the national, city, and local levels of affected countries should make sustained efforts to combat traffic-related air pollution by effective implementation and enforcement of policies. Of immerse importance, targets should be placed on reducing the most important source of atmospheric pollution in cities, which is motor vehicle-related emissions. In this regard, efforts should be made to promote environmentally sustainable transport and transport systems. Lowering emissions is one of the major means of improving local air quality, in addition to being the most frequent approach to coping with vehicular pollution. First and foremost, measures should be taken to improve or change the manner in which people get around as the most cost-effective means of reducing vehicle emissions [28]. These include restriction policies such as emission standards, mandatory vehicle inspections, technology and fuel improvement (electric vehicles, biofuels, and natural gas), restrictions on privately owned vehicles, integrated public transportation, rapid transit, and promotion of active transport such as proper bike and walking lanes [29–30]. The use of plug-in hybrid electric vehicles was also found to achieve greater outcomes in terms of pollutant reduction as well as reduce greenhouse gas [31]. Currently, the USA and China are the largest markets for plug-in hybrid electric vehicles, which are relatively less popular in the Southeast Asia region due to high cost, technical limitations of electric cars, and charging inconveniences [32]. There is a need for transport policies to encourage electric vehicle use with price incentives together with investment in more readily available recharging stations [32].

Active transportation (walking and bicycling) has been mentioned in many studies as the best way to reduce car congestion and lower vehicle emissions while also having important health co-benefits, in particular through increased physical activity [33]. Previous studies have quantified the health benefits of replacing car trips with active transportation trips in urban areas [34, 35]. However, considering the risk of exposure to environmental air pollutants and safety while on the road, the implementation of active transport should integrate a network of streets with bike lanes and pedestrian priority at intersections to make active transport trips convenient, pleasant, and safe [35]. Secondly, along with the promotion of active transport, urban planners and policy makers should have a low emission zone to minimize the risk of exposure to atmospheric pollution among active transport users. Heavy traffic coupled with high levels of traffic air pollution levels along active transport routes may pose a major obstacle to its use.

Another important solution to reducing atmospheric air pollution is absorbing existing pollution by increasing vegetation coverage. Numerous studies reported the capacity of urban trees and shrubs to serve as biological filters and mitigate air pollution [36]. Vegetation can serve as a sink for atmospheric particulate matter and is an interface that can absorb organic matter, chemicals, and heavy metals that adhere to particulate matter [37]. Therefore, urban landscape planning and management to increase vegetation and canopy areas along the sides of busy roads may effectively decrease traffic-related pollutants and benefit commuters.

In addition, there is a need to equip the air quality regulatory authorities with effective and efficient urban air quality management plans, primarily to help maintain urban air quality within the prescribed limits or standards. It has been reported that urban air quality management within countries plays in important role in air quality monitoring and emission inventory [4]. However, this remains a challenge in many developing countries where urban air quality management plans are either in process of development or do not exist at all [4]. The ultimate role of urban air quality management planning is essentially to ensure that the impact of air pollution on the local population remains minimal [4]. It is therefore particularly important for local governments to develop effective air quality management plans for managing urban air pollution in a sustainable and effective long-term manner. Policy makers and urban planners should reinforce stringent air pollution control policies to reduce traffic emission of hazardous air, promote the use of active transport, and enhance green spaces in the cities. More importantly, stringent monitoring and supervision are needed to improve enforcement of traffic emission control. There is also a need for literacy policies and interventions to increase commuter literacy in air pollution and equip them with knowledge of effective pollution control strategies.

1.5. Best practices to reduce personal exposure

While national policies to reduce air pollution at the source are clearly more effective, it is well established that individual-level exposure prevention is effective in reducing exposure and health risks. Using a respiratory filter mask offers the most convenient way to reduce inhalation of pollutants. There is mounting evidence to suggest that the use of a respiratory filter mask is beneficial in minimizing the impact of atmospheric air pollution during rush hour traffic. A respiratory filter mask filters particulate matter out of the air and prevents it from entering the respiratory system. If commuting by public transport is inevitable, wearing a respiratory filter mask is the most effective way to reduce inhalation of particles and mitigate their negative health effects. A study showed that wearing even a simple inexpensive face mask has the potential to protect susceptible individuals and prevent cardiovascular events in cities with high concentrations of ambient air pollution and provides an alternative that may lead to reduced cardiovascular morbidity and mortality [38]. It has also been reported that reduction in symptoms and improvement in cardiovascular health were observed when patients with coronary heart disease were protected from exposure to particulate air pollution by the use of highly efficient face mask (e.g., N95 equivalent) [39]. Wearing a face mask was also found to be beneficial for people without cardiovascular health problems. Among healthy subjects, wearing a face mask appears to abrogate the adverse effects of air pollution on blood pressure and heart rate variability and to decrease cardiovascular risk [38].

Apart from face mask protection, a large body of evidence demonstrates that a healthy diet including fruits and vegetables and supplement intake may protect against the air pollution-induced health threat. Omega-3 polyunsaturated fatty acids (PUFAs) from fish oil were found to prevent the negative impact of $PM_{2.5}$ on heart rate variability [40] and protect against the deleterious cardiac and lipid effects induced by acute exposure to particulate matter [41]. In another study, health outcomes of $PM_{2.5}$, including heart rate variability, were modified by dietary intake of micronutrients (folate, vitamin B6 and B12, methionine) [42]. Antioxidant supplementation (vitamins E and C) was found to be helpful in reducing oxidative stress in the body associated with airborne contamination [43]. Thus, being in a polluted air environment, it is critical that commuters should have a healthy diet with adequate intake of essential micronutrients to prevent the development of chronic diseases, particularly cardiovascular and pulmonary diseases. Increased intake of antioxidants, as well as other anti-inflammatory nutrients, is also important to reduce air pollution-induced oxidative stress and inflammation, particularly among those with cardiovascular disease, asthma, and other chronic inflammatory diseases [44]. Further, it has been suggested that, in addition to a healthy diet, drinking additional water can reduce throat irritation and help the kidneys flush out any absorbed toxins [45, 46].

As public transport commuters are exposed to pollution more than car commuters because they are in the open air during commuter waiting or walking along the road, carpooling as a way of transport may reduce the time spent outside the busy road. Of note, several barriers to carpooling have been reported. Inconveniences of sharing, sense of privacy, self-confidence, and self-gratification relating to driving "my own car" may serve as barriers to carpooling with privately owned vehicles [47]. To be functional, this system of transport should be promoted to elicit general awareness about private interests in carpooling to reduce health hazards, reduce traffic and pollution, and save costs. The current Uber and GrabCar concepts greatly reduce pollution and traffic congestion. Nevertheless, carpooling in Uber and GrabCar usually addresses the one person per vehicle problem with a two person per vehicle solution. It would be better if Uber and GrabCar extended to more than two persons per vehicle to fully maximize the concept of carpooling.

2. Physical and psychological health impacts on public transport commuters: evidence in Malaysia

In Malaysia, a swelling urban population and increased column of motorized traffic in cities have resulted in severe air pollution affecting the surrounding city environment as well as the health of people in the cities. The number of commuters and commute durations has increased substantially in recent decades. People in the city are spending long hours traveling to and from work. Most commuters are exposed to vehicle emission during transit (**Figure 1**). Considering that air pollution caused by traffic is the scourge of many modern cities worldwide and likewise in Malaysia, we conducted a study to examine the health impact of commuting and the mitigation measures practiced by public transport commuters. Presently, the understanding of the impact of air pollution from congestion on roads on the health of people in Malaysia who are exposed during their daily commute is very limited. Such a study is highly warranted and will be used to inform policy making related to traffic and air quality management and mitigation intervention in terms of health effects.

From June to October 2016, a total of 800 public transport commuters in urban cities in the heart of Selangor, Malaysia, were interviewed face to face. The commuters were approached at the Light Rail Transit train stations along the Kelana Jaya and Ampang lines in the state of Selangor, Malaysia. In the interview, self-reported adverse health effects (both physical health, 15 items, and psychological health, 7 items) associated with exposure to atmospheric air pollutants during the daily commute were queried. Self-reported control measures used by participants to mitigate their exposure to atmospheric air pollutants were also assessed.

Figure 1. Commuters exposed to vehicle emission during transit.

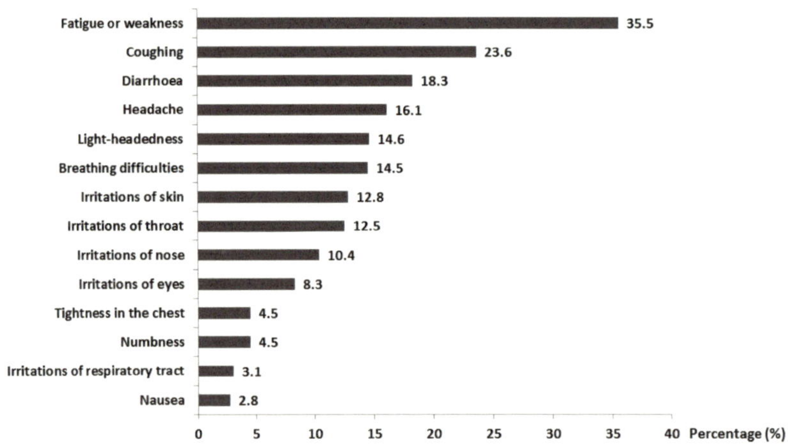

Figure 2. Proportion of adverse physical health effects experienced (%) during the daily commute.

The majority (56.5%) of respondents were aged 30 years and below (mean age, 30.8 years; standard deviation, 6.6; age range, 18–54 years). The ethnic and gender distribution of the study participants closely matched that of the general Malaysian population. A total of 41% of the study respondents spent more than 1 hour commuting by public transport daily. When the respondents were queried on adverse physical health effects they experienced from exposure to air pollution during their daily commute, as shown in **Figure 2**, the highest percentage reported physical fatigue or weakness (35.5%), followed by coughing (23.6%). A considerable proportion reported headache (16.1%), light-headedness (14.6%), and breathing difficulties (14.5%). Air pollution has a wide range of effects on human health. This study evidenced the experience of various physical health impacts commonly associated with exposure to atmospheric air pollutants by daily public transport users.

The physical effects of air pollution on human health has been studied and reported more than the psychological health impacts. However, the impact of air pollution on psychological health is equally prominent. In an animal model, exposure to exhaust emissions increased anxiety- and depression-like behavior and led to impaired memory in rats [48]. Few studies reported psychological health impacts in Western countries [49, 50]; however, psychological health impacts of being exposed to vehicular traffic-related pollutants have rarely been evident in the Asian region. In this study, as shown in **Figure 3**, the most common psychological impact associated with exposure to air pollution during the daily commute was stress (13.0%), followed by insomnia (5.0%) and feeling moody and anxious (3.6%). Depression was reported by 1.1% of respondents. Although the proportion of psychological health effects experienced are lower than that of physical health effects, mental and emotional well-being is essential to overall health. The proportions, although low, have significant clinical implications.

In this study females and the older age group reported higher levels of overall adverse physical and psychological health impacts. However, there were no differences in adverse health outcomes by income and educational levels. As the study is cross-sectional and cannot be used to infer causality, future longitudinal studies are needed to confirm the causal relationship between exposure to road traffic pollution and health impacts. Furthermore, this study recorded self-reported assessment of symptoms and thus was subject to self-reporting bias.

Figure 4 shows the proportion of control measures undertaken by respondents to reduce exposure to atmospheric pollutants during their daily commute. Despite exposure to atmospheric air pollutants during their daily commute, the vast majority of the study participants reported never or rarely practicing control measures to alleviate the adverse impact of air pollution during their daily commute. Relatively higher proportions reported practicing healthcare measures such as drinking more water and increasing their consumption of fruits and vegetables rather than using respiratory protection. On a positive note, consuming a healthy diet and drinking water may help to improve the body's immune system and alleviate the adverse impacts of air pollution [46]. Although face mask protection is widely used by many commuters in many cities worldwide and regarded as the most common and effective means of reducing inhalation of atmospheric air pollutants, the majority of the study

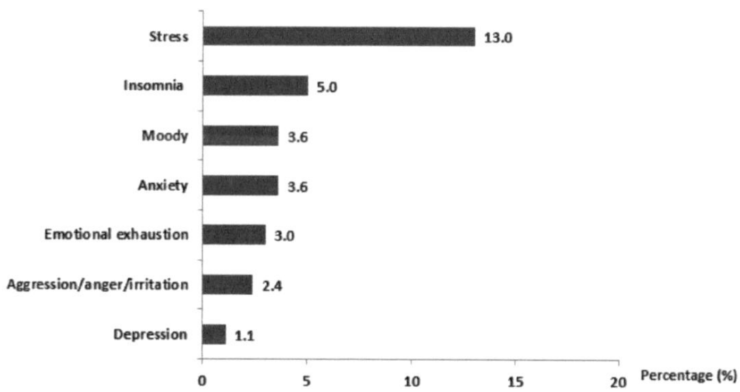

Figure 3. Proportion of the sample reporting adverse psychological health effects experienced during the daily commute.

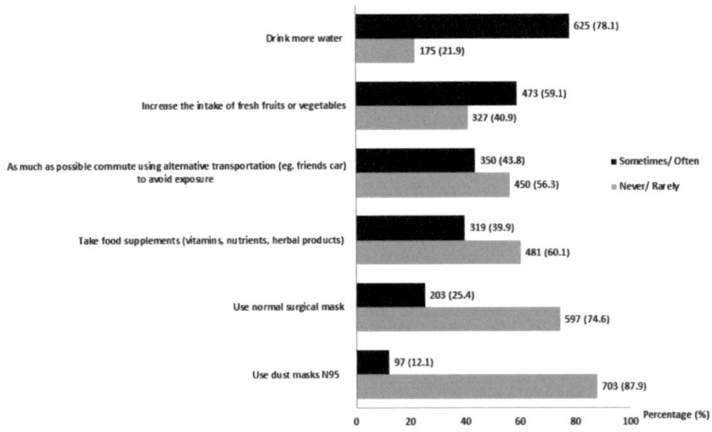

Figure 4. Proportion of respondents reporting the use of control measures during daily commuting.

respondents (74.6%, n = 597) reported never or rarely using a face mask. Face mask protection is not common among commuters in Malaysia and is rarely seen among the public, even on smoke-engulfed busy streets. Facial mask in Malaysia is inexpensive and readily available in most pharmacy outlets. There is a need to know if there is a feeling of embarrassment among the Malaysian public at being seen wearing a facial protection masks as face mask use in the community is not widespread (**Figure 5**), or reluctance to use a face mask is due to the perception of discomfort along with a feeling of difficulty breathing in hot and humid weather. In Japan, China, and Hong Kong, wearing a surgical mask is a social norm [51]. The use of a facial mask in tropical countries of Southeast Asia, such as Indonesia, India, and Malaysia, is relatively less common. If face mask usage becomes widespread and commonly seen, perhaps more people will be likely to use them. In this regard, more publicity is needed to encourage widespread use of face mask protection against atmospheric air pollutants in cities. Therefore, reluctance or barriers to the use of face masks during the daily commute or while on streets with polluted air are issues that need further investigation. The use of facial masks should be encouraged in Southeast Asian countries.

This study also found poor usage of N95 face protection. Although the use of simple surgical mask protection is recommended for commuters and has been found to have beneficial cardiovascular effects of reducing exposure to particulate air pollutants [38], the use of N95 face protection is more effective. Particles smaller than 2.5 μm are widespread in vehicle emissions and absorbed into the bloodstream, initiating the pro-inflammatory cascade, which leads to many adverse health effects. The N95 is a US government-certified mask that blocks 95% of particulate matter (PM) smaller than 2.5 μm—or even smaller, at 0.3 μm. Emission of $PM_{2.5}$ from road vehicles is an important source of atmospheric air pollution in cities. Consequently, the levels of $PM_{2.5}$ near urban roadsides are consistently high [52]; therefore, the use of N95 will ideally offer better protection. In Malaysia, the prevalence of normal surgical mask use is low; therefore, enhancement of use of N95 may require extra effort. The public, especially daily commuters, should be made aware that wearing an N95 mask is more useful in avoiding the detrimental effects of ambient air pollutants [53] and is especially recommended for

Figure 5. Wearing a mask to reduce inhalation of air pollutant is not a common sight in Malaysia.

people with chronic respiratory disease [54]. In a separate report, we compared the use of a facial protection mask and an N95 mask among commuters in the event of severe haze, and increases in the use of both facial protection masks and N95 masks were reported [55]. Nevertheless, the proportion of the population that uses N95 masks is still low: only 66.0% of respondents reported using N95 masks while commuting during severe haze [55]. The haze phenomenon in Malaysia is often due to illegal burning of forests and peat and forest fires from neighboring countries. In the event of severe haze, respiratory protection is required when going outdoors and especially among commuters who are already exposed to vehicle emissions. While a face mask may be effective at filtering large particles, a face mask, by design, does not filter small particles in the air such as vehicle emissions and airborne particles. In contrast, an N95 is a respiratory protective device designed to achieve a close facial fit and is more efficient in the filtration of small particles. Therefore, using an N95 respirator which provides a good fit and high filtration should be highly recommended to public transport commuters.

The majority of respondents reported never or rarely trying to commute using alternative transportation as much as possible. These findings imply the considerable need to encourage as much as possible the use of alternative transit mode options while traveling around or to the city to work. Carpooling may help reduce contact with atmospheric air pollutions and

		Total physical and psychological health experience score during daily commute days		
		Score 0–4 (n = 684)	Score 5–22 (n = 116)	P value
Sociodemographic data				
Age group (years old)				
30 and below	452 (56.5)	387 (85.6)	65 (14.4)	
31–40	265 (33.1)	230 (86.8)	35 (13.2)	0.389
>40	83 (10.4)	67 (80.7)	16 (19.3)	
Gender				
Male	424 (53.0)	370 (87.3)	54 (12.7)	0.159
Female	376 (47.0)	314 (83.5)	62 (16.5)	
Marital status				
Single	414 (51.8)	356 (86.0)	58 (14.0)	0.689
Married	386 (48.3)	328 (85.0)	58 (15.0)	
Ethnicity				
Malay	388 (48.5)	332 (85.6)	56 (14.4)	
Chinese	271 (33.9)	235 (86.7)	36 (13.3)	0.636
Indian	138 (17.3)	115 (83.3)	23 (16.7)	
Others	3 (0.4)	2 (66.7)	1 (33.3)	
Highest education level				
Secondary level and below	146 (18.2)	124 (84.9)	22 (15.1)	0.796
Tertiary level	654 (81.8)	560 (85.6)	94 (14.4)	
Occupation				
Professional and managerial	419 (52.4)	358 (85.4)	61 (14.6)	
Skilled/non-skilled worker	314 (40.6)	278 (85.5)	47 (14.5)	0.951
Student	54 (6.8)	46 (85.2)	8 (14.8)	
Housewife	2 (0.2)	2 (100.0)	0 (0.0)	
Monthly income (MYR)				
5000 and below	651 (81.4)	559 (85.9)	36 (14.1)	
>5000	149 (18.6)	125 (83.9)	56 (14.1)	0.826
Other risks				
Chronic diseases				
Yes	51 (6.4)	40 (78.4)	92 (14.1)	0.521
No	749 (93.6)	644 (86.0)	24 (16.1)	

		Total physical and psychological health experience score during daily commute days		
		Score 0–4 (n = 684)	Score 5–22 (n = 116)	P value
Smoking status				
Yes	309 (38.6)	266 (86.1)	43 (13.9)	0.758
No	491 (61.4)	418 (85.1)	73 (14.9)	
Living near a highway				
Yes	292 (36.5)	255 (87.3)	37 (12.7)	0.297
No	508 (63.5)	429 (84.4)	79 (15.6)	
Duration spent commuting by public transport (hours)				
1 hour and below	472 (59.0)	412 (87.3)	60 (12.7)	0.102
>1 hour	328 (41.0)	272 (82.9)	56 (17.1)	
Control measures				
Total control measure score				
Score 0–8	440 (55.0)	371 (84.3)	69 (15.7)	0.314
Score 9–18	360 (45.0)	313 (86.9)	47 (13.1)	

Table 1. Factors associated with total physical and psychological health experience score (n = 800).

should be encouraged. As indicated above, carpooling by Uber and GrabCar that allow more than two persons per vehicle would reduce the chance of exposure while being out on the road or waiting at stops or stations while minimizing number of private cars on the road.

As shown in **Table 1**, further investigation of the association between control measures and overall physical and psychological health impacts found that respondents with lower use of control measures experience a higher total health impact. It was also found that those who spend longer commuting were more likely to report higher overall physical and psychological health impacts. However, none of the associations were statistically significant; therefore, further studies are needed to verify these associations. Of note, the absence of significant differences between control measures and health impact found in this study could also imply the need to enhance control measures among people who experience a higher health impact. Ideally those who experience a higher health impact should practice greater mitigation measures than those with lower health impacts. Consequently, our findings may imply that commuters in the city of Kuala Lumpur, in particular those who experience a higher health impact, should be encouraged to carry out greater mitigation practices against exposure to atmospheric air pollution during the daily commute. We also found higher health impact score among those with longer commuting duration, although the association is not statistically significant. Further studies are needed to confirm this finding. Of note, this is a cross-sectional study; therefore, the cause and effect relationship between control measures and health impact cannot be established.

3. Directions for future research

It is globally recognized that battling traffic congestion, spending a long time commuting, and exposure to atmospheric pollutants during the daily commute result in serious adverse effects on the health and general well-being of commuters. There is a paucity of epidemiological studies regarding the adverse health impacts of atmospheric air pollution, which is worsened by traffic-related air pollution in rapidly developing cities. Although commuting undeniably has a significant impact on public health outcomes, in the Asian region in particular, little empirical evidence has documented the adverse health effects associated with commuting. More research, especially in the form of longitudinal studies, is critically needed to provide insights into the population health impact of traffic-related air pollution in Asia, the world's most rapidly developing region.

Given the many factors such as vehicles, fuel type, topography, meteorological conditions, and a complex array of health determinants such as social, behavioral, and lifestyle factors that may influence the degree of atmospheric pollution and its subsequent impact on human health, emphasis is warranted across all countries on tailoring their health impact assessments according to the specific needs of the affected populations within cities. Measuring local specific health outcomes associated with using public transport would enable specific tailored monitoring and evaluation of and accountability for the health of the relevant population. Subsequently, action plans must be tailored to individual cities and countries at large.

Likewise, the assessment of mitigation practices must also be tailored to understanding barriers to carry out specialized prevention in order to determine target areas for education in the respective cities and countries. It is also important to assess how commuters in a specific area respond to the adverse effects of commuting to work and exposure to traffic pollutants daily. It is vital to assess the level of preventive measures carried out by commuters as well as the types of preventive measures undertaken and the barriers to their use. Such information varies among cities across Asia as well as in Western countries and may be influenced by various factors such as culture, religion, and climate, as well as policy and governance. Identifying the specific health impact of air pollution and level of mitigation measures taken provides an important basis for customized environmental governance by governments of the respective countries.

There is also a dire need to investigate ways to improve upon the measurement of ambient air quality, provide accurate air quality data, and effectively inform the public when ambient air quality is at a dangerous level. Coupled with that, publicly available air quality data can support communication efforts so that people can act to protect themselves against exposure. Furthermore, an informed public can also help to advocate for prevention initiatives such as a reduction in the use of private vehicles to decrease traffic volume during peak-period traffic. As the health effects of air pollution can be reduced via improved understanding and subsequent proper self-preventive measures, a study to identify knowledge gaps and barriers to reducing exposure to traffic-related air pollution during commuting should be carried out to identify intervention strategies, especially for the susceptible population of public transport commuters.

http://dx.doi.org/10.5772/intechopen.79694

Acknowledgements

The study was funded by Grand Challenge-SUS (Sustainability Science) Grants GC002C-15SUS and GC002A-15SUS.

Conflict of interest

The authors declare no conflict of interest.

Author details

Li Ping Wong[1]*, Haridah Alias[1], Nasrin Aghamohammadi[2], Nik Meriam Nik Sulaiman[3], Hualiang Lin[4] and Jinshun Zhao[5]

*Address all correspondence to: wonglp@ummc.edu.my

1 Julius Centre University of Malaya (JCUM), Department of Social and Preventive Medicine, Faculty of Medicine, University of Malaya, Kuala Lumpur, Malaysia

2 Centre for Occupational and Environmental Health, Department of Social and Preventive Medicine, Faculty of Medicine, University of Malaya, Kuala Lumpur, Malaysia

3 Department of Chemical Engineering, Faculty of Engineering, University of Malaya, Kuala Lumpur, Malaysia

4 Department of Medical Statistics and Epidemiology, School of Public Health, Sun Yat-sen University, Guangzhou, China

5 Department of Preventive Medicine, Zhejiang Provincial Key Laboratory of Pathological and Physiological Technology, Medical School of Ningbo University, Ningbo, Zhejiang, China

References

[1] Rivas I, Kumar P, Hagen-Zanker A. Exposure to air pollutants during commuting in London: Are there inequalities among different socio-economic groups? Environment International. 2017;**101**:143-157

[2] Baldasano JM, Valera E, Jimenez P. Air quality data from large cities. The Science of the Total Environment. 2003;**307**:141-165

[3] Martin J. The European environment: State and outlook 2010: Synthesis. European Environment Agency. 2010:31-33

[4] Gulia S, Nagendra SS, Khare M, Khanna I. Urban air quality management-a review. Atmospheric Pollution Research. 2015;**6**:286-304

[5] Zhang K, Batterman S. Air pollution and health risks due to vehicle traffic. The Science of the Total Environment. 2013;**450**:307-316

[6] Brook RD, Rajagopalan S, Pope CA, Brook JR, Bhatnagar A, Diez-Roux AV, Holguin F, Hong Y, Luepker RV, Mittleman MA, Peters A. Particulate matter air pollution and cardiovascular disease: An update to the scientific statement from the American Heart Association. Circulation. 2010;**121**:2331-2378

[7] Cheng J. WHO: Air Pollution has Become the World's Largest Environmental Health Risk. Available from: http://health.huanqiu.com/health_news/2015-06/6581336.html?referer=huanqiu. [Accessed: 16-05-2018]

[8] WHO. 7 million deaths annually linked to air pollution. Central European Journal of Public Health. 2014;**22**:53-59

[9] Van Donkelaar A, Martin RV, Brauer M, Kahn R, Levy R, Verduzco C, Villeneuve PJ. Global estimates of ambient fine particulate matter concentrations from satellite-based aerosol optical depth: Development and application. Environmental Health Perspectives. 2010;**118**:847

[10] Adar SD, Davey M, Sullivan JR, Compher M, Szpiro A, Liu LJ. Predicting airborne particle levels aboard Washington state school buses. Atmospheric Environment. 2008; **42**:7590-7599

[11] McCreanor J, Cullinan P, Nieuwenhuijsen MJ, Stewart-Evans J, Malliarou E, Jarup L, Harrington R, Svartengren M, Han IK, Ohman-Strickland P, Chung KF. Respiratory effects of exposure to diesel traffic in persons with asthma. The New England Journal of Medicine. 2007;**357**:2348-2358

[12] Peters A, Von Klot S, Heier M, Trentinaglia I, Hörmann A, Wichmann HE, Löwel H. Exposure to traffic and the onset of myocardial infarction. The New England Journal of Medicine. 2004;**351**:1721-1730

[13] Shima M, Nitta Y, Adachi M. Traffic-related air pollution and respiratory symptoms in children living along trunk roads in Chiba prefecture, Japan. Journal of Epidemiology. 2003;**13**:108-119

[14] Brauer M, Hoek G, Van Vliet P, Meliefste K, Fischer PH, Wijga A, Koopman LP, Neijens HJ, Gerritsen J, Kerkhof M, Heinrich J. Air pollution from traffic and the development of respiratory infections and asthmatic and allergic symptoms in children. American Journal of Respiratory and Critical Care Medicine. 2002;**166**:1092-1098

[15] Ji D, Zhang J, He J, Wang X, Pang B, Liu Z, Wang L, Wang Y. Characteristics of atmospheric organic and elemental carbon aerosols in urban Beijing, China. Atmospheric Environment. 2016;**125**:293-306

[16] Sun C, Luo Y, Li J. Urban traffic infrastructure investment and air pollution: Evidence from the 83 cities in China. Journal of Cleaner Production. 2018 Jan 20;**172**:488-496

[17] Yang G, Wang Y, Zeng Y, Gao GF, Liang X, Zhou M, Wan X, Yu S, Jiang Y, Naghavi M, Vos T. Rapid health transition in China, 1990-2010: Findings from the global burden of disease study 2010. The Lancet. 2013;**381**:1987-2015

[18] Chen Z, Wang JN, Ma GX, Zhang YS. China tackles the health effects of air pollution. The Lancet. 2013;**382**:1959-1960

[19] Resosudarmo BP, Napitupulu L. Health and economic impact of air pollution in Jakarta. The Economic Record. 2004;**80**:65-66

[20] Matus K, Nam KM, Selin NE, Lamsal LN, Reilly JM, Paltsev S. Health damages from air pollution in China. Global Environmental Change. 2012;**22**:55-66

[21] Quah E, Boon TL. The economic cost of particulate air pollution on health in Singapore. Journal of Asian Economics. 2003;**14**:73-90

[22] Abd Manan N, Abdul Manaf MR, Hod R. The Malaysia haze and its health economy impact. A literature review. Malaysian Journal of Public Health Medicine. 2018;**18**:1-9

[23] Gray KM. From content knowledge to community change: A review of representations of environmental health literacy. International Journal of Environmental Research and Public Health. 2018;**15**:466

[24] Qian X, Xu G, Li L, Shen Y, He T, Liang Y, Yang Z, Zhou WW, Xu J. Knowledge and perceptions of air pollution in Ningbo, China. BMC Public Health. 2016;**16**:1138

[25] Liu X, Zhu H, Hu Y, Feng S, Chu Y, Wu Y, Wang C, Zhang Y, Yuan Z, Lu Y. Public's health risk awareness on urban air pollution in Chinese megacities: The cases of shanghai, Wuhan and Nanchang. International Journal of Environmental Research and Public Health. 2016;**13**:845

[26] Haque MS, Singh RB. Air pollution and human health in Kolkata, India: A case study. Climate. 2017;**5**:77

[27] De Pretto L, Acreman S, Ashfold MJ, Mohankumar SK, Campos-Arceiz A. The link between knowledge, attitudes and practices in relation to atmospheric haze pollution in peninsular Malaysia. PLoS One. 2015;**10**:e0143655

[28] Wright L, Fulton L. Climate change mitigation and transport in developing nations. Transport Reviews. 2005;**25**:691-717

[29] Pojani D, Stead D. Sustainable urban transport in the developing world: Beyond megacities. Sustainability. 2015;**7**:7784-7805

[30] Slovic AD, de Oliveira MA, Biehl J, Ribeiro H. How can urban policies improve air quality and help mitigate global climate change: A systematic mapping review. Journal of Urban Health. 2016;**93**:73-95

[31] Kantor I, Fowler MW, Hajimiragha A, Elkamel A. Air quality and environmental impacts of alternative vehicle technologies in Ontario, Canada. International Journal of Hydrogen Energy. 2010;**35**:5145-5153

[32] Delang CO, Cheng WT. Hong Kong people's attitudes towards electric cars. International Journal of Electric and Hybrid Vehicles. 2013;**5**:15-27

[33] Macmillan A, Connor J, Witten K, Kearns R, Rees D, Woodward A. The societal costs and benefits of commuter bicycling: Simulating the effects of specific policies using system dynamics modeling. Environmental Health Perspectives. 2014;**122**:335-344

[34] Rojas-Rueda D, De Nazelle A, Teixidó O, Nieuwenhuijsen MJ. Replacing car trips by increasing bike and public transport in the greater Barcelona metropolitan area: A health impact assessment study. Environment International. 2012;**49**:100-109

[35] Rojas-Rueda D, De Nazelle A, Andersen ZJ, Braun-Fahrländer C, Bruha J, Bruhova-Foltynova H, Desqueyroux H, Praznoczy C, Ragettli MS, Tainio M, Nieuwenhuijsen MJ. Health impacts of active transportation in Europe. PLoS One. 2016;**11**:e0149990

[36] Brugge D, Patton AP, Bob A, Reisner E, Lowe L, Bright OJ, Durant JL, Newman J, Zamore W. Developing community-level policy and practice to reduce traffic-related air pollution exposure. Environmental Justice. 2015;**8**:95-104

[37] Tallis M, Taylor G, Sinnett D, Freer-Smith P. Estimating the removal of atmospheric particulate pollution by the urban tree canopy of London, under current and future environments. Landscape and Urban Planning. 2011;**103**:129-138

[38] Langrish JP, Mills NL, Chan JK, Leseman DL, Aitken RJ, Fokkens PH, Cassee FR, Li J, Donaldson K, Newby DE, Jiang L. Beneficial cardiovascular effects of reducing exposure to particulate air pollution with a simple facemask. Particle and Fibre Toxicology. 2009;**6**:8

[39] Langrish JP, Li X, Wang S, Lee MM, Barnes GD, Miller MR, Cassee FR, Boon NA, Donaldson K, Li J, Li L. Reducing personal exposure to particulate air pollution improves cardiovascular health in patients with coronary heart disease. Environmental Health Perspectives. 2012;**120**:367

[40] Romieu I, Téllez-Rojo MM, Lazo M, Manzano-Patiño A, Cortez-Lugo M, Julien P, Bélanger MC, Hernandez-Avila M, Holguin F. Omega-3 fatty acid prevents heart rate variability reductions associated with particulate matter. American Journal of Respiratory and Critical Care Medicine. 2005;**172**:1534-1540

[41] Tong H, Rappold AG, Diaz-Sanchez D, Steck SE, Berntsen J, Cascio WE, Devlin RB, Samet JM. Omega-3 fatty acid supplementation appears to attenuate particulate air pollution–induced cardiac effects and lipid changes in healthy middle-aged adults. Environmental Health Perspectives. 2012;**120**:952

[42] Baccarelli A, Cassano PA, Litonjua A, Park SK, Suh H, Sparrow D, Vokonas P, Schwartz J. Cardiac autonomic dysfunction: Effects from particulate air pollution and protection by dietary methyl nutrients and metabolic polymorphisms. Circulation. 2008;**117**:1802-1809

[43] Possamai FP, Júnior SÁ, Parisotto EB, Moratelli AM, Inácio DB, Garlet TR, Dal-Pizzol F, Wilhelm Filho D. Antioxidant intervention compensates oxidative stress in blood of subjects exposed to emissions from a coal electric-power plant in South Brazil. Environmental Toxicology and Pharmacology. 2010;**30**:175-180

[44] Péter S, Holguin F, Wood LG, Clougherty JE, Raederstorff D, Antal M, Weber P, Eggersdorfer M. Nutritional solutions to reduce risks of negative health impacts of air pollution. Nutrients. 2015;**7**:10398-10416

[45] Cai DP, He YM. Daily lifestyles in the fog and haze weather. Journal of Thoracic Disease. 2016;**8**:E75

[46] Rajper SA, Ullah S, Li Z. Exposure to air pollution and self-reported effects on Chinese students: A case study of 13 megacities. PLoS One. 2018;**13**:e0194364

[47] Galizzi MM. The economics of car-pooling: A survey for Europe. In: Paper for the Workshop: Highways-Cost and Regulation in Europe. Bergamo, Italy: Universita degli Studi di Bergamo; 2004

[48] Salvi A, Patki G, Liu H, Salim S. Psychological impact of vehicle exhaust exposure: Insights from an animal model. Scientific Reports. 2017;**7**:8306

[49] Shin J, Park JY, Choi J. Long-term exposure to ambient air pollutants and mental health status: A nationwide population-based cross-sectional study. PLoS One. 2018;**13**:e0195607

[50] Sass V, Kravitz-Wirtz N, Karceski SM, Hajat A, Crowder K, Takeuchi D. The effects of air pollution on individual psychological distress. Health & Place. 2017;**48**:72-79

[51] Wada K, Oka-Ezoe K, Smith DR. Wearing face masks in public during the influenza season may reflect other positive hygiene practices in Japan. BMC Public Health. 2012;**12**:1065

[52] Ginzburg H, Liu X, Baker M, Shreeve R, Jayanty RK, Campbell D, Zielinska B. Monitoring study of the near-road PM2. 5 concentrations in Maryland. Journal of the Air & Waste Management Association. 2015;**65**:1062-1071

[53] Zhao P, Yu KP, Lin CC. Risk assessment of inhalation exposure to polycyclic aromatic hydrocarbons in Taiwanese workers at night markets. International Archives of Occupational and Environmental Health. 2011;**84**:231-237

[54] Jiang XQ, Mei XD, Feng D. Air pollution and chronic airway diseases: What should people know and do? Journal of Thoracic Disease. 2016;**8**:E31

[55] Wong LP, Alias H, Aghamohammadi N, Ghadimi A, Sulaiman NM. Control measures and health effects of air pollution: A survey among public transportation commuters in Malaysia. Sustainability. 2017;**9**:1616

Air Pollution in an Urban Area of Mexico: Sources of Emission (Vehicular, Natural, Industrial, and Brick Production)

Valeria Ojeda-Castillo, Sergio Alonso-Romero,
Leonel Hernández-Mena,
Paz Elizabeth Álvarez-Chávez and
Jorge del Real-Olvera

Additional information is available at the end of the chapter

http://dx.doi.org/10.5772/intechopen.80000

Abstract

In recent years, the interest in aerosol particles has increased due to concerns about the effects on human health. The study of the chemical characterization (organic matter, sulfates, nitrates, and black carbon) has improved the knowledge about the negative contribution of chemicals to the environment. The identification of secondary processes (from pollutants such as SO_2, NO_x, and PAHs) and their role when combined with environmental factors such as humidity, solar radiation, and temperature are also of interest. With this background, this chapter seeks to highlight the most recent findings on the chemical composition of aerosol particles in the ambient air of one of the main cities of Mexico: the Metropolitan Area of Guadalajara. This megalopolis has almost 60% of the population of the Jalisco state, approximately 2.2 million vehicles and an extensive artisan brick production. Furthermore, due to its geographical position, it experiences frequent episodes of thermal inversion and exposition to high levels of solar radiation, mainly during the first half of the year.

Keywords: emissions, particle composition, sources, secondary formation, human health

1. Introduction

This chapter briefly describes the results about specific studies of the concentration of particulate pollutants in ambient air and on the presence of several associated chemical species in the Guadalajara Metropolitan Area (GMA), the second most important city of Mexico due to its territorial extension and the size of its population. The particulate material implies the coexistence of hundreds and even thousands of chemical compounds with different physical, chemical, and biological properties; responsible in part for the numerous adverse effects on the environment and the health of the human population. These effects are the main reasons that motivate the study of particles in this and other cities of the world. Specifically, those particles less than or equal to 2.5 microns ($PM_{2.5}$) are considered as a criterion contaminant since they determine the quality of the ambient air. Therefore, the information presented here focuses mainly on data of $PM_{2.5}$ concentrations and chemical components, such as trace metals and toxic elements, anions and cations, black carbon, and polycyclic aromatic hydrocarbons (PAHs), oxy-PAHs, and quinones. However, only a few studies show results about particles of other sizes, such as those smaller than or equal to 10 μm (PM_{10}), particles less than or equal to 1 micron (PM_1) (which are not regulated in Mexico), and the chemical composition of organic compounds such as PAHs and derived quinones. Moreover, the study and analysis of chemical species could insinuate the potential sources of emission (direct or indirect, natural or anthropogenic) and also indicate the existence of processes of secondary formation of particles in the ambient air (as in the case of nitrates, sulfates, and quinones). Studies on the chemical characterization of particles in the GMA have been carried out mainly in two sites inside of the city, Centro and Miravalle. In this context, this chapter addresses the most significant findings regarding the knowledge of GMA particles in the ambient air, as well as some aspects of its chemical composition and how the latter may suggest its origin.

2. Aerosols

2.1. Definition and properties

Aerosols are a collection of solid or liquid particles suspended in a gas (also called particulate material and defined by the acronym PM). The term aerosol refers to "aerosol particles" without considering the suspension gas [1]. The aerosols formed by particles of different sizes (among them TSP, PM_{10}, $PM_{2.5}$, and PM_1, as well as ultrafine particles), together with other gases in the ambient air, in some cases are considered criteria pollutants and define the quality of the air in a city or town [2].

The aerosols come from numerous natural and anthropogenic sources, such as the burning of biomass, incomplete combustion of fossil fuels, volcanic eruptions, soil and mineral dust, sea salt and biological materials (plant fragments, microorganisms, pollen, etc.). In addition, aerosols can be formed in the atmosphere (secondary aerosol) by the conversion from a gas to particles (e.g., through mechanisms such as nucleation and condensation) and by heterogeneous and multiphase chemical reactions [3].

Among the properties of aerosols that negatively influence the health of the human population are the aerodynamic diameter and the chemical composition of the particles [4]. The size of the PM ranges from micrometer to nanometer, and those with a diameter smaller than 10 μm (PM_{10}) could enter the respiratory tract reaching the lungs.

The chemical composition of the particles is extremely complex and has a wide range of chemical species (inorganic acids and salts, metals, water, and a mixture of organic compounds of low volatility) in variable concentrations. Given that coarse, fine, and ultrafine particles are generated through different processes, it is expected that each of these groups of PM has a unique composition of chemical species, which allows establishing inferences about the sources and training processes. Because the chemical composition of PMs has adverse environmental and human health effects, research is still underway intensively to clarify the molecular composition of PM chemical species [5].

As an illustrative example, typical elements found in the Earth's crust may contribute to the chemical composition of coarse particles [6]. Sulfates (derived from SO_2 gas) and organic compounds are prominent compounds present in ultrafine particles due to their formation by homogeneous nucleation. In addition to inorganic species such as nitrates (derived from NO_x) and ammonium, the chemical composition of fine particles also includes sulfates and organic polar species, obtained from the coagulation of ultrafine particles. The formation of fine particles through the combustion and condensation processes of low volatility vapors also results in particular combustion products such as organic carbon (between 20 and 90% of the mass of particles in the troposphere [7]), elemental carbon, and traces of metals. In this way, these two properties together cause particles to serve as carriers of toxic substances into the body, a situation that associates them with different health problems in human populations around the world.

It should be mentioned that the physical and chemical diversity of PM particles manifests itself in different ways. The particle sizes vary in different orders of magnitude, which strongly influences the life of the particles in the atmosphere and therefore, in the spatial extent of their influence. The shape and morphology of the particles are also highly variable and may comprise spheres, crystalline, or irregular fragments, needles, agglomerates, and dendritic entities. The individual particles may be chemically uniform or contain chemically different surrounding and nuclear materials, while a set of particles may comprise similar particles or a diversity of geometries [8].

2.2. Human health and environmental effects

The PM has a significant adverse impact on human health. The World Health Organization (WHO) estimates that in 2016, 95% of the world's population lived in places where pollution levels are higher than the standards of health protection over air quality, that the ambient air pollution in cities and rural areas caused 4.1 million deaths worldwide, and the loss of 106 million years of human disability that same year [9]. Since PMs are a critical component of the air whose presence defines its quality, the International Agency for Research on Cancer (IARC) has classified PMs in its Group 1 (carcinogenic to humans), together with agents such as arsenic and asbestos [10].

Recent investigations have reported that exposure to $PM_{2.5}$ may also contribute to the incidence and development of diabetes mellitus, premature birth, low birth weight, and postneonatal infant mortality [11]; moreover, low neurodevelopment [12, 13] and the diagnosis of autism [14].

About the environmental impacts, the low visibility (an optical indicator of air quality) is attributed mainly to the dispersion and absorption of visible light by particles in the atmosphere [15] with diameters mostly between 0.1 and 1.0 μm. The fine particles scatter visible light much more effectively per unit mass since their size is comparable to the wavelength of visible light [16]. Additionally, damage to ecosystems can also be mentioned that can result in the reduction of biodiversity and loss of environmental goods and services [17].

3. Studies of aerosol in the GMA

3.1. Concentration of PM

The average concentrations of $PM_{2.5}$ in Centro and Miravalle exceeded the annual limits during the periods of January–December 2007, January–June 2008, and April 2009–January 2010 (as a whole the three periods represent 27 months of sampling, justifying its comparison with the annual limit of the standard of 15 μg m^{-3}). In Miravalle, concentrations of $PM_{2.5}$ were higher during 2007, probably due to higher industrial activity and the natural barriers in that area that prevent the evacuation of pollutants; in Centro and Miravalle were between 2.9 and 3.5 times higher than the annual limit of the standard, respectively. At the Miravalle site, the higher levels of PM were evident in the dry seasons (warm and cold) when compared to the ones measured in the rainy season. In general, during the cold period, particle levels are increased by low temperatures, unfavorable conditions for the dispersion of pollutants, by the increase in the consumption of fossil fuels, and because the conversion of gases to particles is favored [18]. The lowest temperatures were recorded in February 2008; they occurred particularly in Miravalle, and it is likely that the processes already mentioned were related to the high concentrations of $PM_{2.5}$.

Other studies considered $PM_{2.5}$ measurements at shorter sampling periods and the values contrasted against the standard limit for a period of 24 h (65 μg m^{-3}, [19]). Only sites like Miravalle in May–June 2009, and Centro and Tlaquepaque in March–May 2014, showed concentrations close to this limit, but not exceeding it. It should be mentioned that the levels of this same particle size in the studies carried out at the end of 2014 were compared with the 24 h standard of 45 μg m^{-3} [19] without exceeding the limit in any case (**Figure 1**).

PM_{10} levels in the area of the Environmental Fragility Polygon (POFA, by the acronym in Spanish) exceeded 1.6 times the standard for 24 h (120 μg m^{-3}), which suggests a risk for the health of the local population. Likewise, concentrations of PM_{10} were 4–10 times higher than those of $PM_{2.5}$ (in their daily levels), indicating the domain of emission sources of coarse particles. The low correlation between fine and coarse particles suggests different emission sources (**Table 1**).

Another environmental factor, the rain, contributes to the temporal variation of the particles, which are grounded by direct drag or because air humidity induces their growth in size and precipitation. Some hygroscopic compounds favor this phenomenon in the particles, such as

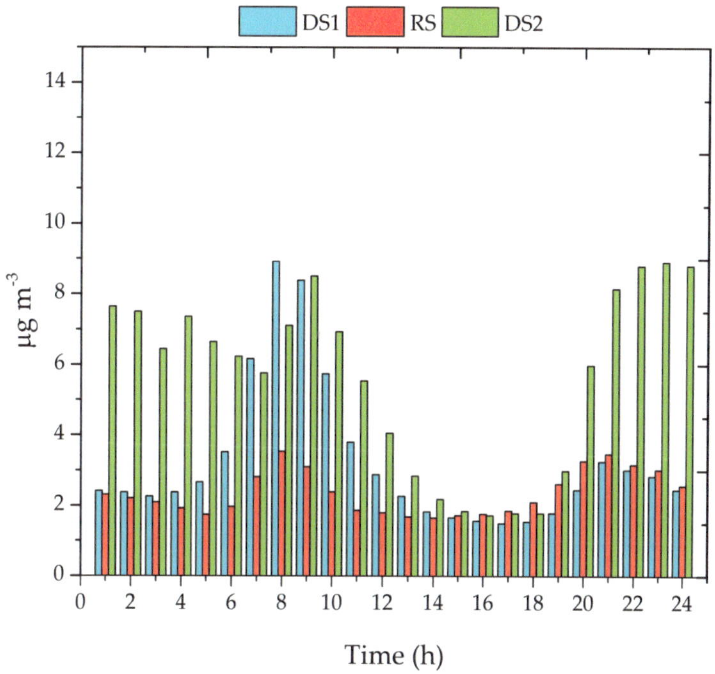

Figure 1. Map of monitoring stations (gray circles) into GMA, the red circles indicate the sampling sites.

sulfates and nitrates [33]. For the 2009–2010 period, Centro and Miravalle sites did not show differences in their particle concentrations; only in Miravalle were observed higher levels during the cold dry season against the rainy season.

Other studies considered $PM_{2.5}$ measurements at shorter sampling periods and the values contrasted against the standard limit for a period of 24 h (65 μg m^{-3}, [34]). Only sites like Miravalle in May–June 2009, and Centro and Tlaquepaque in March–May 2014, showed concentrations close to this limit, but not exceeding it. It should be mentioned that the levels of this same particle size in the studies carried out at the end of 2014 were compared with the 24 h standard of 45 μg m^{-3} [19] without exceeding the limit in any case.

PM_{10} levels in the area of the Environmental Fragility Polygon (POFA, by the acronym in Spanish) exceeded 1.6 times the standard for 24 h (120 μg m^{-3}), which suggests a risk for the health of the local population. Likewise, concentrations of PM_{10} were 4–10 times higher than those of $PM_{2.5}$ (in their daily levels), indicating the domain of emission sources of coarse particles. The low correlation between fine and coarse particles suggests different emission sources.

Regarding PM_1 levels, there is no Mexican regulation. It is impossible to conclude about the impacts as done before. However, Centro and Tlaquepaque sites seem to have equal PM_1 concentrations, despite the characteristics of each of them concerning the types of emission sources, so that further studies on this particle size will be needed in other parts of the city to establish relationships. It is important to note that the concentrations of PM_1 are recent and the first for this type of pollutants in the GMA.

Sampling period	Sites	Size	Average ($\mu g\ m^{-3}$)	Chemical compounds	NOM-025-SSA1 1993 and 2014
Jan–Dec, 2007	Centro	$PM_{2.5}$	44.10	Heavy metals and toxic elements[20]; anions & cations [21]; black carbon** [22]	
	Miravalle		52.80		
Jan–Dec, 2008	Centro	$PM_{2.5}$	59.09	Anions & cations [23]; heavy metals and toxic elements [22], black carbon*** [24]	
	Miravalle		81.36		
May 25th-June 6th, 2009	Centro	$PM_{2.5}$	39.30	Heavy metals and toxic elements; anions & cations [25].	NOM-025-SSA1-1993
	Miravallle		58.00		
	Centro	$PM_{2.5}$	21.00	Heavy metals and toxic elements; anions & cations; black carbon [26]	PM_{10}
	Miravalle		37.00		120 $\mu g\ m^{-3}$ (24 h)
	AUG		21.00		50 $\mu g\ m^{-3}$ (annual)
	Centro	PM_{10}	46.00		$PM_{2.5}$
	Miravalle		79.00		65 $\mu g\ m^{-3}$ (24 h)
	AUG		43.00		15 $\mu g\ m^{-3}$ (annual)
Apr 2009–Jan 2010	Centro	$PM_{2.5}$	33.00[1,2]	Oxy-PAHs [27]	NOM-025-SSA1-2014
	Miravalle		45.00[1,2]		PM_{10}
					75 $\mu g\ m^{-3}$ (24 h)
					40 $\mu g\ m^{-3}$ (annual)
					$PM_{2.5}$
					45 $\mu g\ m^{-3}$ (24 h)
					12 $\mu g\ m^{-3}$ (annual)
Apr 6th–18th, 2013	POFA*	PM_{10}	187.83	Anions & cations**** [28]	
		$PM_{2.5}$	29.33		
Mar–May, 2014	Centro	$PM_{2.5}$	57.07[1,2]	PAHs & quinones [29, 30]	
July–Sep, 2014	Tlaquepaque		61.68[1,2]		
	Centro		28.97[1,2]		
	Las Águilas		28.23[1,2]		
Aug 24th–30th, 2014	Oblatos	$PM_{2.5}$	22.67	Black carbon & VOCs [31]	
	Loma Dorada		15.67		
	Las Pintas		30.00		
	Las Águilas		17.33		
	Centro		17.00		
Apr–June, 2015	Centro	PM_1	30.92[2]	PAHs & quinones [32]	
	Tlaquepaque		21.80[2]		

[1]Corresponds to the medians of the sampling periods.

[2]in ng m^{-3}.

*Acronym in Spanish.

**For this study, the sampling sites were Centro and CIATEJ.

***Corresponds to the period of January–August, 2008.

****The chemical species mentioned were only determined in PM_{10}.

Table 1. GMA aerosols research.

3.2. Heavy metals and toxic elements

The objective of these studies was also to generate reliable information about the possible sources of emission into the atmosphere, essential to establish actions that could reduce current levels of pollution in the ambient air and to even be able to estimate the degree of exposure of the human population. From the study of $PM_{2.5}$ (2007) and its elemental composition in Centro and Miravalle, emission sources were established through the estimation of the enrichment factors (EF).

The EF must be understood as a relation of the elements of interest with an abundant element of the soil (usually Fe and K), but also present in the particles. An EF > 5 suggests an anthropogenic source (different to the geological or natural origin). In Centro and Miravalle, Pb, Cd, Cr, Cu, Mo, Sb, Se, Tl, and Zn, as well as Ni and V in Centro had high FE, suggesting that they came from anthropogenic sources. On the other hand, Co, Fe, Mg, Ca, Mn, Sr, and Ti, clearly showed evidence from resuspension of dust either by the winds or by vehicular traffic (natural origin).

For the study of January–June 2008, the most abundant elements in both sites were Fe, Ti, Zn, Mg, and Pb. Fe, Ti, and Mg showed FE values that indicate a geological origin, while Zn and Pb probably are incorporated into the atmosphere by vehicle emissions and the use of these elements in non-ferrous alloys [35, 36]. The elemental composition profile of this study was related to the $PM_{2.5}$ elementary profiles of the international USEPA SPECIATE database (version 4.2). This database gathers scientific data from the profiles of specific particle sources. The profiles of Miravalle and Centro coincided with the profiles of Paved Road Dust (PRD), Diesel Exhaust (DE) and industrial soil (IS) ($r^2 = 94\%$, $r^2 = 72\%$, $r^2 = 68\%$, and $r^2 = 89\%$, $r^2 = 75\%$, $r^2 = 66\%$, respectively). Miravalle, in addition to the high industrial activity, has an intense flow of heavy vehicles which can be the origin of the resuspension of particles, phenomenon increased by large land areas without vegetation during the dry-season months. In Centro, the coincidence with the profiles of PRD and IS could be understood by the phenomenon of particle resuspension, or by the transport of emissions from nearby places with some industrial activity. In both sites, the DE profile was influenced by public transportation that works with diesel.

During 2009, some trends were consistent with those observed in 2008 regarding the abundance of those elements related to $PM_{2.5}$. It is highlighted that the Fe contributed 60 and 72% to the total elemental mass in Centro and Miravalle, respectively. In this way, it can be seen how different emission sources, whether natural or anthropogenic, could affect the composition of the $PM_{2.5}$ particles, and at the same time give indications of the specific emission sources.

3.3. Anions and cations

From the study in 2009, sulfate, nitrate, and ammonium were the more abundant compounds in Centro and Miravalle; sulfate and calcium showed high concentrations in Miravalle. In the case of sulfate, this is a derivate from SO_2 which comes from direct industrial emissions and the burning of fossil fuels with sulfur content. While nitrates and ammonium are formed from NO_x and NH_3 (ammonia), the latter gas is emitted mainly by the vehicle fleet that uses fuels such as gasoline and diesel in the GMA [37, 38]. These results suggest that the relatively constant concentrations of sulfate and nitrate in both sites during the year are because the levels emitted from parent gases are similar. For sulfates during winter, for example, the precursor gas is at high concentrations, while during the summer, the high temperature and high solar radiation favor the oxidation of the precursor and increase the formation of sulfate [39]. This study was complemented by an ion balance, which established the existence of neutralization processes between the chemical species of the particles. For Centro, the presence of slightly acidic particles was determined, while in Miravalle, the particles showed an alkaline nature. The acid form of sulfates can influence the bioavailability of some metals, increase the toxicity of the particles [40], augment the adsorption of compounds into the particles, or create hygroscopic environments in them, all of which favor the solubility of some gases, promoting that they can reach deep areas of the respiratory tract [41], increasing the deposition of toxic compounds in the lungs. In the case of alkalinity, the potential effects on human health from exposure to particles with this property are now not very clear [42].

On the other hand, during 2008, sulfate, nitrate, and ammonium compounds showed the highest levels, without differences between sites, suggesting sources and formation processes in common, particularly in the southwest of the GMA. In Centro, the ion mass represented 46% of the mass of the $PM_{2.5}$, while in Miravalle, they did not exceed 28%. Studies like [43] indicate that the mass of these three ion species contributed with 40% and up to 65% to the mass of $PM_{2.5}$. The results on the contribution of the major inorganic ions to $PM_{2.5}$ can serve as a basis for establishing the connection between fine particles and possible adverse impacts on human health.

The short study on ion species also in 2009 in the two sites mentioned above, coincide in sulfate, nitrate, and ammonium as those of higher abundance in the previous works, but also highlight the presence of calcium and potassium. An impressive result was to find higher levels of nitrates concerning sulfates and ammonium. In Miravalle, nitrate represented 68% of the mass of total ions, while in Centro it was 57%. The abundance of these ions shows the critical role that secondary sources have in the formation of fine particles.

The last study about species of ions and particles (PM_{10}) was carried out on three different sites within the POFA area (on specific days of April 2013) [28]. In this area pollution problems are recurrent in the soil, water, and air because it has changed land use, environmental deterioration, loss of its original fauna and flora. The low air quality due to particles, ozone, and the presence of unpleasant odors is the most severe environmental problem in the POFA. Species such as sulfate, nitrate, chloride, calcium, sodium, and ammonium were the most abundant, without differences in their concentrations between sites (except chloride), which suggest the probable homogeneity of their levels in the area of study (**Figure 2**).

The total ions contributed with 15% of the PM_{10} mass (other components such as organic matter, elemental carbon, and mineral silica could integrate the particles). Calcium and sulfate showed significant correlations with PM_{10} ($r = 0.70$ and $r = 0.64$, respectively), evidence of emission sources of coarse particles in the POFA. Also, chloride could derive from local industry, burning biomass, and burning of agricultural waste. Finally, the PM_{10} collected in this study had an acidic nature; this is based on the ratio of total sulfate and total ammonium.

3.4. Black carbon

Black carbon (BC), or elemental carbon, comes from incomplete combustion of diesel vehicles, electric power plants and various industrial processes, wood burning and forest fires. BC plays an important role due to its porous nature and absorption, as well as in global climate

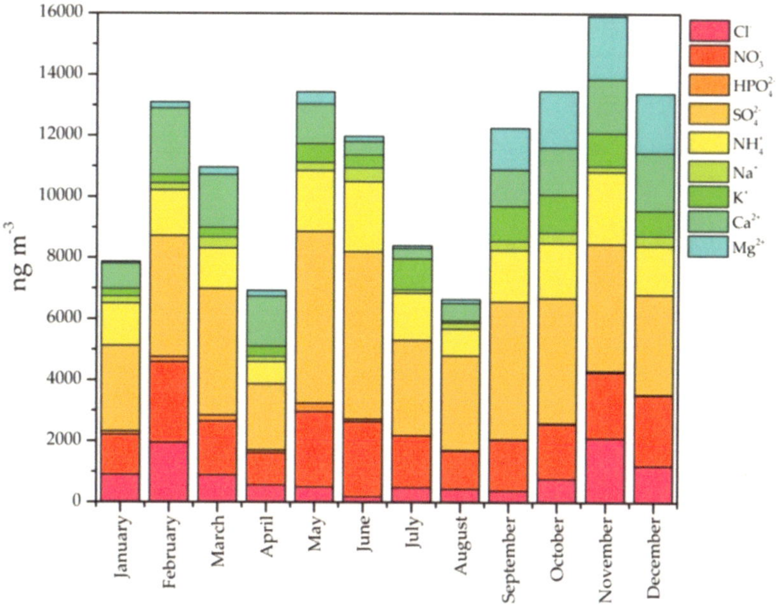

Figure 2. Average monthly concentration of different species of anions and cations associated with $PM_{2.5}$ of Miravalle during 2007. Sulfate, nitrate, and ammonium are the species that contribute most to the concentrations of fine particles throughout the year.

change by affecting the intensity of solar radiation [44]. Due to the porosity of the BC particles and their enormous surface area, they can adsorb significant amounts of mutagenic and carcinogenic organic compounds [45]. Therefore, BC is a risk to human health due to its ability to carry carcinogens to lungs [46].

During 2007, BC associated with $PM_{2.5}$ was monitored in Centro and CIATEJ [22]. Overall, low temperatures and unfavorable conditions for the dispersion of pollutants resulted in high concentrations at both sites (November and December). During July and August, a significant reduction of this pollutant was observed (factor 1:0.5 or 1:0.3). The presence of humidity and the intensity of the winds that generate vertical movements of air masses may also favor the BC dispersion, altogether with the frequency and severity of rainfall.

Furthermore, BC concentrations during 24 h periods were studied. In the morning, between 8 and 9, the highest level of concentration was observed during all months of the year (the magnitude of this peak is high or low, depending on the months of the dry or rainy season). This phenomenon occurred in both places, but with higher levels in Centro. During evenings, both sites showed a second-high level of BC concentrations, and this can be seen between 20 and 21 h. With some exceptions, the concentrations of this last event do not exceed the levels observed during the morning. **Figure 3** shows the trends and variations of BC concentrations throughout the day.

The abovementioned results are due to the increased population activities during the mornings, mainly because of the displacement from home to job, which implies a higher use of the public transport and private vehicles. Around midday not only the intensity of this activity of the population is reduced, but also better environmental conditions for the dispersion of

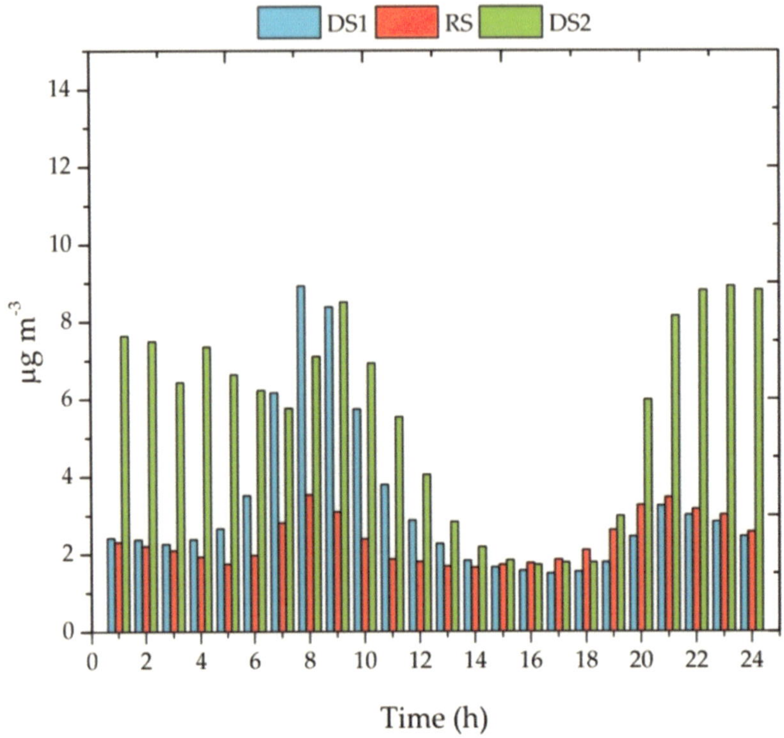

Figure 3. Hourly variation of BC concentrations per season in the Centro site during 2008. DS1—corresponds from January to May; RS—corresponds from June to October; DS2—corresponds from November to December.

pollutants are favored, thus generating lower concentrations of BC. The increase in BC at night is due to vehicles circulation related to people returning home after workday.

The BC data from the Centro could be correlated with the environmental concentrations of criteria pollutants measured at the same site. It is highlighted that with CO, NO_2, and SO_2 the correlations were high, but not in the case of ozone (a contaminant of secondary origin). These results support the primary origin of BC (mainly diesel vehicles). The second study on BC was conducted between January and August 2008, in addition to being carried out along with a manual sampling of $PM_{2.5}$; both pollutants were monitored and sampled at Centro and Miravalle sites. The results confirm some of the trends described for the previous year of monitoring (variations throughout the day and time when the concentration increase due to peaks in population activities).

An interesting contribution of this last study was the estimation of the contribution that BC makes to the $PM_{2.5}$, less than 10%. This value is similar to that estimated in other cities such as City of Mexico, Santiago de Chile, Helsinki, among others [47–49]. In Centro and Miravalle, a moderate relation was found between the concentrations of BC and CO (carbon monoxide) and confirms a common source of emission (burning of biomass and petroleum fuels).

Regarding BC emissions, those sources that use organic matter (burning biomass) as fuel during simple combustion processes and that lack of technological equipment to control their emissions (gases and particles), could be responsible for a large part of the pollution problems in urban areas of our country; this is the case of artisan brick producers. In Mexico, Jalisco is the second state with the largest number of artisan brick producers, 2500 for 2012, 46% of which are in the GMA. The artisan brick industry contributes significantly to the emission of particulate ($PM_{2.5}$ and PM_{10}) and gaseous pollutants such as CO, CO_2, NH_3, among others. It should be noted that a significant fraction of these particles, especially $PM_{2.5}$, corresponds to BC [50], another relevant reason for their study. The scarce technology and standardization in the artisan brick production is a constant not only locally, but also in the country. Although there are numerous variants of kiln types designed for this productive activity (such as the MK2 one), the most common are the field campaign kiln and the fixed-walls kiln [51]. Few of these kilns (including the MK2 one) have additional emission control systems to reduce the gases and emitted particles. In this sense, it is possible to minimize particles dragging by water curtains, being one of the most straightforward and most efficient principles [52], although it does not act in the same way as the emission gases (the absorption of gases is sensitive to the increase in the temperature of the washing water). Our research group has observed the behavior of particles removal during the process of brick cooking when using recirculated washing water in an MK2 kiln, finding increasing water turbidity with time. The change in turbidity and its level (583 ± 16.7, FTU) is related to the number of total solids (827 ± 62.38, mg L^{-1}) therein which are removed by the washing water, thus contributing to the efficiency of reducing emissions to the atmosphere [53]. As part of these observations, a significant increase in the temperature of the washing water was also measured (up to 52°C), as well as a decrease in pH (3.45) due to the absorption of part of the CO_2 emissions (the most abundant gas). Both of the washing water properties change with time up to the final limit removal of CO_2, CO, and NO not exceeding 26, 37, and 22%, respectively. The washing water of this process could experience a physical-chemical treatment due to the ratio BOD/COD = 0.17 because the BOD level is low compared to other types of wastewater.

3.5. Organic components as emission indicators (polycyclic aromatic hydrocarbons and oxy-PAHs)

As regards the organic matter associated with the particles, only a small percentage of groups of compounds have been characterized, such as polycyclic aromatic hydrocarbons (PAHs), oxygenated polycyclic aromatic hydrocarbons (oxy-PAHs), among others [54]. The identification and quantification of PAHs, a diverse group of complex compounds with two or more fused benzene rings, has been intensively investigated [55]. In the atmosphere, PAHs can be present in particles (adsorb/absorb on their surface) or in the gas phase. It is expected that high molecular weight PAHs can be found mainly in the fine particles [56] because they have a more upper specific surface area and therefore greater adsorption than coarse particles [57].

The United States Environmental Protection Agency (USEPA) has defined 16 PAHs as a priority due to the known carcinogenic and mutagenic properties of them. It is important to note that only the European Union considers the regulation of PAHs in ambient air expressed in benzo[a]pyrene concentration (the carcinogen of reference by the IARC) at levels of 1 ng m^{-3} per year [58].

An essential aspect of PAHs is that they can be decomposed by photodegradation, chemical reactions with others pollutants in the urban environment, or gas-particle partition.

Photodegradation is the central mechanism of chemical decomposition of 4–6 ring PAHs associated with particles. While the processes of degradation of PAHs in the gas phase involve reactions with hydroxyl radicals ($^{\bullet}OH$), ozone (O_3), nitrate radicals (NO_3), and reactions of basic species with acids [59]. On the other hand, in the atmosphere, the oxidation of PAHs through homogeneous and heterogeneous reactions leads to the formation of oxy- and nitro-PAHs [60, 61]. These last species are also emitted with PAHs during incomplete combustion processes [62, 63]. Oxy- and nitro-PAHs are potentially more mutagenic than PAHs, and some of these substances are furthermore potentially carcinogenic [64, 65].

PAHs are typical components of gasoline and diesel. They are mainly emitted into the ambient air by exhaust gases from vehicles and other sources of incomplete combustion (approximately 90% of total PAHs from these sources) [59, 66]. PAH emissions can also arise from tobacco smoke and biomass burning [67, 68]. Knowledge of the sources, chemistry, and fate of each PAHs is crucial to propose successful prevention and mitigation, thus reducing the exposure of the population [69].

A study developed in the GMA evaluated the environmental levels of 14 PAHs and 14 oxy-PAHs, associated with $PM_{2.5}$ in Centro and Miravalle. Three seasons from April-2009 to January-2010 were investigated [27]. In general, the highest concentrations occurred during the cold dry season, due to the steady trend of organic pollutants to bind to particles; this behavior is directly influenced by metrological phenomena such as thermal inversions (presented 78% of the days of the year) and the stagnation of pollutants due to the orographic characteristics of the area. The most abundant compounds were benzo[a]pyrene, indene[123-*cd*]pyrene and benzo[*ghi*] perylene; the former contributing to 75% of the total carcinogenic potential and representing a public health risk in addition to the concentration of respirable $PM_{2.5}$. Regarding sources of PAHs, indene[123-*cd*] pyrene and benzo[*ghi*]perylene can be considered as markers for gasoline [70]. About oxy-PAHs, the most abundant were perinaphtenone, 2-methylanthracenedione, 7*H*-benzo[*de*]anthracene-7one, and 9,10-anthraquinone. The correlations between oxy-PAHs and criteria pollutants suggest that they come from combustion processes, specifically the burning of gasoline or diesel; also, there are indications of contribution by secondary transformation sources.

Barradas-Gimate et al. [29] carried out a study in 2014 at the Centro and Tlaquepaque during the warm dry season and at Centro and Las Aguilas during the rainy season, with the objective of determining the atmospheric levels and sources of quinones in $PM_{2.5}$. The highest concentration of Σ_{16}PAHs occurred in Tlaquepaque (7.62 ± 2.03 ng m^{-3}) in the warm dry season and the lowest was recorded in Las Aguilas (2.98 ± 0.55 ng m^{-3}). During the rainy season, since Tlaquepaque has the characteristics of a site for the reception of pollutants concerning the Centro station, it can be inferred that there is an influence of the masses of air coming from the emission sites that promote higher concentrations of these pollutants. In this study compounds such as benzo[*ghi*] perylene, indene [*123*-cd] pyrene, benzo[*a*]pyrene and benzo[*b*] fluoranthene, 1,2-benzanthraquinone, followed by 9,10-anthraquinone and 5,12-naphtacenoquinone were the most abundant. It is suggested that some of the compounds may be associated with diesel particles and heavy vehicle activity. In general, the quinones indicate having a primary origin (vehicular). However, evidence of secondary formation was found for 9,10-phenanthrenequinone, a compound toxic for humans [71].

Similarly, Ojeda-Castillo et al. [32] carried out a simultaneous sampling of gas phase and PM_1 for the 16 priority PAHs and 8 quinones during the warm dry season of 2015 in the Centro and Tlaquepaque, being the first time that this particle size is measured in conjunction with the organic gas phase in the GMA. The levels of Σ_{16}PAHs (sum of gas phase and PM_1) were found in the range of 69.34–209.77 ng m^{-3}. The most abundant compounds in the gas phase were naphthalene and acenaphthylene, while in PM_1 was benzo[b]fluoranthene and chrysene. The ratios of PAHs confirmed that these compounds come from the combustion of diesel. It is important to note that in previous studies, this trend is followed regardless of the particle size, which reaffirms the need to take actions to mitigate vehicle emissions in the city.

Additionally, the health risk of lung cancer is calculated by exposure of PAHs to the respirable aerosol at a level of 1.7×10^{-3}. It is considered that population exposure to PAHs lower than 10^{-6} is acceptable [72, 73]. Therefore, this result could indicate a potential risk to health. More research is needed on this topic to reduce uncertainties and to specify the evaluation by vulnerable groups and exposure times. Regarding to other chemicals, the environmental levels of Σ_8Quinones were found in the range of 1.36–12.23 ng m^{-3}. The most abundant quinone in both phases was 1,4-phenanthrenequinone, which has been reported as a direct emission marker [74]. In the gas phase, quinones represent about 58% of the collected emissions. Inference results from sources with criteria pollutants and meteorological parameters indicate primary sources as the main contribution of quinones in PM_1, and secondary formation of high molecular weight quinones in the gas phase and PM_1.

4. Conclusions

For the GMA, the analysis of the chemical composition of particulate pollutants, especially $PM_{2.5}$, has allowed us to have reliable information on some of the most common sources of emissions that affect the ambient air and probably the health of its population. The geological origin of some of the components of the particles, or an anthropogenic origin such as diesel or gasoline emissions, is one of the main sources of emission of heavy metals and toxic elements of the particles. At the same time, the most abundant species such as sulfates, nitrates, and ammonium allow us to understand that there are chemical compounds that are formed in the atmosphere and end up being incorporated into the mass of the particles. The parent gases of these secondary pollutants are frequently emitted directly from the combustion processes, highlighting the importance of this source. In the case of BC in the particles, its porous absorption nature makes them a relevant component in the issue of human health, since it is not only its abundance greater in the smallest particles, but also particulate matter reaches deeper regions of the respiratory tract, bringing numerous chemical compounds into the body, a situation that could be the origin of multiple respiratory diseases. It is probable that in some areas of the GMA one of the most critical sources of BC is the artisanal bricks industry. Thus, particular attention and monitoring must be given to the measurement of their emissions, as well as to public policies for modernization of its production processes. As for PAHs, oxy-PAHs, and quinones, the results suggest that the predominance comes from gasoline or diesel burning emissions, that is, direct emissions. It is emphasized that for the PM_1 the quinones come from primary sources, while a secondary origin is attributed to the high molecular weight of quinones in the gas phase and the PM_1.

Although there have been advances in the knowledge of the origin of some of the components of the particles of different sizes in the ambient air of the GMA, the studies have focused only on a few groups of chemical compounds, and in a few places; therefore, for most of the city pollution information is not really known. In this way, direct efforts of local authorities and research groups trying to understand the particulate pollutants processes in the GMA are still needed.

Acknowledgements

The authors would like to thank SENER and CONACYT (Project FSE-2013-2105-234633 and No. 183444, respectively) for financial support.

Author details

Valeria Ojeda-Castillo[1]*, Sergio Alonso-Romero[2], Leonel Hernández-Mena[1], Paz Elizabeth Álvarez-Chávez[1] and Jorge del Real-Olvera[1]

*Address all correspondence to: ojeda.c.v@gmail.com

1 Center for Research and Assistance in Technology and Design of the State of Jalisco (CIATEJ, A.C.), Guadalajara, Jalisco, México

2 Center of Applied Innovation in Competitive Technologies (CIATEC, A.C.), León, Guanajuato, México

References

[1] Pöschl U. Atmospheric aerosols: Composition, transformation, climate and health effects. Angewandte Chemie, International Edition. 2005;**44**(46):7520-7540

[2] Fuzzi S, Baltensperger U, Carslaw K, Decesari S, Denier Van Der Gon H, Facchini MC, et al. Particulate matter, air quality and climate: Lessons learned and future needs. Atmospheric Chemistry and Physics. 2015;**15**(14):8217-8299

[3] Moise T, Flores JM, Rudich Y. Optical properties of secondary organic aerosols and their changes by chemical processes. Chemical Reviews. 2015;**115**(10):4400-4439

[4] Harrison RM, Yin J. Particulate matter in the atmosphere: Which particle properties are important for its effects on health? Science of the Total Environment. 2000;**249**:85-101

[5] Bell ML. Assessment of the Health Impacts of Particulate Matter Characteristics [Internet]. Health Effects Institute. 2012. Available from: http://www.ncbi.nlm.nih.gov/pubmed/22393584

[6] Finlayson-Pitts BJ, Pitts JJN. Chemistry of the Upper and Lower Atmosphere. Theory, Experiments, and Applications. New York, USA: Academic Press; 2000. p. 969

[7] Kanakidou M, Seinfeld JH, Pandis SN, Barnes I, Dentener FJ, Facchini MC, et al. Organic aerosol and global climate modelling: A review. Atmospheric Chemistry and Physics. 2005;**5**:1053-1123

[8] Heal MR, Kumar P, Harrison RM. Particles, air quality, policy and health. Chemical Society Reviews. 2012;**41**:6606

[9] Institute Health Effects. State of Global Air 2018. Special Report. Boston, MA; 2018

[10] IARC. IARC: Outdoor Air Pollution a Leading Environmental Cause of Cancer Deaths [Internet]. Press release. Lyon, France; 2013 [cited 2017 Jul 20]. Available from: https://www.iarc.fr/en/media-centre/iarcnews/pdf/pr221_E.pdf

[11] Feng S, Gao D, Liao F, Zhou F, Wang X. The health effects of ambient $PM_{2.5}$ and potential mechanisms. Ecotoxicology and Environmental Safety. 2016;**128**:67-74. Available from: 10.1016/j.ecoenv.2016.01.030

[12] Levy RJ. Carbon monoxide pollution and neurodevelopment: A public health concern. Neurotoxicology and Teratology. 2015;**49**:31-40. DOI: 10.1016/j.ntt.2015.03.001

[13] Block ML, Elder A, Auten RL, Bilbo SD, Chen H, Chen JC, et al. The outdoor air pollution and brain health workshop. Neurotoxicology. 2012;**33**(5):972-984. DOI: 10.1016/j.neuro.2012.08.014

[14] Vrijheid M, Casas M, Gascon M, Valvi D, Nieuwenhuijsen M. Environmental pollutants and child health—A review of recent concerns. International Journal of Hygiene and Environmental Health. 2016;**219**(4-5):331-342. DOI: 10.1016/j.ijheh.2016.05.001

[15] Chan YC, Simpson RW, Mctainsh GH, Vowles PD, Cohen DD, Bailey GM. Source apportionment of visibility degradation problems in Brisbane (Australia) using the multiple linear regression techniques. Chemist-Analyst. 1999;**33**:3237-3250

[16] McCartney EJ. Optics of the Atmosphere: Scattering by Molecules and Particles. New York: John Wiley and Sons Inc.; 1976. p. 421

[17] Grantz D, Garner JH, Johnson D. Ecological effects of particulate matter. Environment International. 2003;**29**(2-3):213-239. Available from: http://linkinghub.elsevier.com/retrieve/pii/S0160412002001812

[18] Calvo AI, Pont V, Liousse C, Dupré B, Mariscal A, Zouiten C, et al. Chemical composition of urban aerosols in Toulouse, France during CAPITOUL experiment. Meteorology and Atmospheric Physics. 2008;**102**(3-4):307-323

[19] Secretaria de Salud. NORMA Oficial Mexicana NOM-025-SSA1-2014, Salud ambiental. Valores límite permisibles para la concentración de partículas suspendidas PM_{10} y $PM_{2.5}$ en el aire ambiente y criterios para su evaluación. Diario oficial de la Federacion; 2014. p. 53

[20] Saldarriaga-Noreña H, Hernández-Mena L, Ramírez-Muñiz M, Carbajal-Romero P, Cosío-Ramírez R, Esquivel-Hernández B. Characterization of trace metals of risk to human health in airborne particulate matter ($PM_{2.5}$) at two sites in Guadalajara, Mexico.

Journal of Environmental Monitoring. 2009;**11**(4):887. Available from: http://xlink.rsc.org/?DOI=b815747b

[21] Hernández-Mena L, Saldarriaga-Noreña H, Carbajal-Romero P, Murillo-Tovar MA, Limón-Sánchez MT, López-López A. Presence of the most abundant ionic species and their contribution to $PM_{2.5}$ mass, in the city of Guadalajara, Jalisco (Mexico). Bulletin of Environmental Contamination and Toxicology. 2010;**85**(6):632-637

[22] Hernández-Mena L, Saldarriaga-Noreña H, Murillo-Tovar MA, Amador-Muñoz O, López-López A, Waliszewski SM. Determination of black carbon in fine particles using a semi-continuous method at two sites in the city of Guadalajara, Mexico, during 2007. Bulletin of Environmental Contamination and Toxicology. 2011;**87**(3):336-342

[23] Hernández-Mena L, Saldarriaga-Noreña H, Carbajal-Romero P, Cosío-Ramírez R, Esquivel-Hernández B. Ionic species associated with $PM_{2.5}$ in the City of Guadalajara, México during 2007. Environmental Monitoring and Assessment. 2010;**161**(1-4):281-293

[24] Limon-Sanchez MT, Carbajal-Romero P, Hernandez-Mena L, Saldarriaga-Norena H, Lopez-Lopez A, Cosio-Ramirez R. Black carbon in $PM_{2.5}$, data from two urban sites in Guadalajara, Mexico during 2008. Atmospheric Pollution Research. 2011;**2**(3):358-365. DOI: 10.5094/APR.2011.040

[25] Murillo-Tovar M, Saldarriaga-Noreña H, Hernández-Mena L, Campos-Ramos A, Cárdenas-González B, Ospina-Noreña J, et al. Potential Sources of Trace Metals and Ionic Species in $PM_{2.5}$ in Guadalajara. Atmosphere. 2015;**6**(12):1858-1870. Available from: http://www.mdpi.com/2073-4433/6/12/1834

[26] INE-CIATEJ. Estudio preliminar de compuestos tóxicos en aire ambiente en la Zona Metropolitana de Guadalajara; 2009

[27] Flores-Arriola A. Determinación de Hidrocarburos Aromáticos Policíclicos Oxigenados en las Aeropartículas finas de la Zona Metropolitana de Guadalajara. Universidad Autónoma de Guadalajara; 2012

[28] Hernandez-Mena L, Gallardo Valdez J, Díaz-Torres JJ, Villegas García E. Contaminación del aire por partículas (PM_{10}) en el Polígono de Fragilidad Ambiental, Guadalajara, Jalisco. Reaxion, Ciencia y Tecnoligia University. 2017;**2**(4):12-25

[29] Barradas-Gimate A, Murillo-Tovar M, Díaz-Torres J, Hernández-Mena L, Saldarriaga-Noreña H, Delgado-Saborit J, et al. Occurrence and Potential Sources of Quinones Associated with $PM_{2.5}$ in Guadalajara, Mexico. Atmosphere. 2017;**8**(8):140. Available from: http://www.mdpi.com/2073-4433/8/8/140

[30] Barradas-Gimate A, Murillo-Tovar MA, Hernandez-Mena L, Lopez-Lopez A, de Díaz-Torres JJ. Contribution of the atmospheric photochemical process to the aromatic quinones presence in $PM_{2.5}$ in an urban area. In: International Congress Energy and Environmental Engineering and Management. Paris, France; 2015

[31] INECC-CIATEJ. Evaluación de $PM_{2.5}$, compuestos orgánicos volátiles y ozono para definir las medidas de control en la Zona Metropolitana de Guadalajara; 2014

[32] Ojeda-Castillo V, López-López A, Hernández-Mena L, Murillo-Tovar MA, Díaz-Torres J de J, Hernández-Paniagua IY, et al. Atmospheric distribution of PAHs and quinones in the gas and PM 1 phases in the Guadalajara Metropolitan Area, Mexico: Sources and health risk. Atmosphere. 2018;**9**(4):1-21

[33] McMurry PH. A review of atmospheric aerosol measurements. Atmospheric Environment. 2000;**34**:1959-1999

[34] Secretaria de Salud. NORMA Oficial Mexicana NOM-025-SSA1-1993. Salud ambiental. Criterio para evaluar la calidad del aire ambiente, con respecto a las particulas menores de 10 MICRAS (PM_{10}). Valor permisible para la concentracion de particulas menores de 10 MICRAS (PM_{10}) EN; 1993

[35] Dutkiewicz VA, Qureshi S, Husain L, Schwab JJ, Demerjian KL. Elemental composition of $PM_{2.5}$ aerosols in Queens, New York: Evaluation of sources of fine-particle mass. Atmospheric Environment. 2006;**40**(Suppl. 2):347-359

[36] Chen J, Tan M, Li Y, Zheng J, Zhang Y, Shan Z, et al. Characteristics of trace elements and lead isotope ratios in $PM_{2.5}$ from four sites in Shanghai. Journal of Hazardous Materials. 2008;**156**(1-3):36-43

[37] Báez PA, García MR, BMC d T, Padilla HG, Belmont RD, Amador OM, et al. Origin of trace elements and inorganic ions in PM_{10} aerosols to the South of Mexico City. Atmospheric Research. 2007;**85**(1):52-63

[38] Hodgson E, editor. A Textbook of Modern Toxicology. 3rd ed. New Jersey: John Wiley & Sons, Inc.; 2004. 557 p. DOI: 10.1002/0471646776

[39] Lonati G, Giugliano M, Ozgen S. Primary and secondary components of $PM_{2.5}$ in Milan (Italy). Environment International. 2008;**34**(5):665-670

[40] Ghio AJ, Stoneheurner J, McGee JK, Kinsey JS. Sulfate content correlate with iron concentrations in ambient air pollution particles. Inhalation Toxicology. 1999;**11**(4):293-307

[41] Friedlander SK, Yeh EK. The submicron atmospheric aerosol as a carrier of reactive chemical species: Case of peroxides. Applied Occupational and Environmental Hygiene. 1998;**13**(6):416-420. DOI: 10.1080/1047322X.1998.10389566

[42] Chung A, Herner JD, Kleeman MJ. Detection of alkaline ultrafine atmospheric particles at Bakersfield, California. Environmental Science & Technology. 2001;**35**(11):2184-2190

[43] Ito K, Xue N, Thurston G. Spatial variation of $PM_{2.5}$ chemical species and source-apportioned mass concentrations in New York City. Atmospheric Environment. 2004; **38**(31):5269-5282

[44] Hansen J, Sato M, Ruedy R, Lacis A, Oinas V. Global warming in the twenty-first century: An alternative scenario. Proceedings of the National Academy of Sciences. 2000;**97**(18): 9875-9880. DOI: 10.1073/pnas.170278997

[45] Lighty JAS, Veranth JM, Sarofim AF. Combustion aerosols: Factors governing their size and composition and implications to human health. Journal of the Air & Waste Management Association (1995). 2000;**50**(9):1565-1618

[46] Jiang M, Marr LC, Dunlea EJ, Herndon SC, Jayne JT, Kolb CE, et al. Mobile laboratory measurements of black carbon, polycyclic aromatic hydrocarbons and other vehicle emissions in Mexico City. Atmospheric Chemistry and Physics Discussions. 2005;**5**(4):7387-7414. Available from: http://search.proquest.com/docview/288979812?accountid=171201

[47] Chow JC, Watson JG, Crow D, Lowenthal DH, Merrifield T. Comparison of IMPROVE and NIOSH Carbon Measurements. Aerosol Science and Technology. 2001;**34**(1):23-34

[48] Artaxo P, Oyola P, Martinez R. Aerosol composition and source apportionment in Santiago de Chile. Nuclear Instruments and Methods in Physics Research Section B: Beam Interactions with Materials and Atoms. 1999;**150**(1-4):409-416

[49] Viidanoja J, Sillanpää M, Laakia J, Kerminen VM, Hillamo R, Aarnio P, et al. Organic and black carbon in $PM_{2.5}$ and PM_{10}: 1 Year of data from an urban site in Helsinki, Finland. Atmospheric Environment. 2002;**36**(19):3183-3193

[50] Gobierno de Jalisco—Secretaría de Medio Ambiente para el Desarrollo Sustentable y S de MA y RN. PROAire; 2011-2020

[51] SERpro. Diagnóstico Nacional del sector ladrillero artesanal de México-Informe Final; 2012

[52] Instituto Nacional de Ecología. Tecnologías de control de contaminantes procedentes de Fuentes estacionarias; 2002

[53] Alvarez CPE. Tratamiento de agua residual generada durante el lavado de gases del proceso de cocción tradicional de ladrillo rojo. Centro de Investigación y Asistencia en Tecnología y Diseño del Estado de Jalisco; 2018

[54] Rogge WF, Mazurek M, Hildemann LM, Cass GR, BRT S. Quantification of urban organic aerosols at a molecular level: Identification, abundance and seasonal variation. Atmospheric Environment, Part A, General Topics. 1993;**27**(8):1309-1330

[55] Ravindra K, Sokhi R, Grieken, Van R. Atmospheric polycyclic aromatic hydrocarbons: Source attribution, emission factors and regulation. Atmospheric Environment. 2008; **42**(13):2895-2921

[56] Duan J, Bi X, Tan J, Sheng G, Fu J. Seasonal variation on size distribution and concentration of PAHs in Guangzhou City, China. Chemosphere. 2007;**67**(3):614-622

[57] Sheu HL, Lee WJ, Lin SJ, Fang GC, Chang HC, You WC. Particle-bound PAH content in ambient air. Environmental Pollution. 1997;**96**(3):369-382

[58] Unión Europea. Directiva 2004/107/CE del Parlamento Europeo y del Consejo de 15 de diciembre de 2004 relativa al arsénico, el cadmio, el mercurio, el níquel y los hidrocarburos aromáticos policíclicos en el aire ambiente. L23. 2004;**26**(1):2005

[59] Neilson A. The Handbook of Environmental Chemistry: PAHs and Related Compounds. New York: Springer; 1998. p. 431

[60] Atkinson R, Arey J. Mechanisms of the gas-phase reactions of aromatic hydrocarbons and PAHs with OH and NO_3 radicals. Polycyclic Aromatic Compounds. 2007;**27**:15-40

[61] Keyte IJ, Harrison RM, Lammel G. Chemical reactivity and long-range transport potential of polycyclic aromatic hydrocarbons—A review. Chemical Society Reviews. 2013;**42**(24):9333

[62] Chen Y, Zhi G, Feng Y, Tian C, Bi X, Li J, et al. Increase in polycyclic aromatic hydrocarbon (PAH) emissions due to briquetting: A challenge to the coal briquetting policy. Environmental Pollution. 2015;**204**:58-63

[63] Nalin F, Golly B, Besombes JL, Pelletier C, Aujay-Plouzeau R, Verlhac S, et al. Fast oxidation processes from emission to ambient air introduction of aerosol emitted by residential log wood stoves. Atmospheric Environment. 2016;**143**:15-26

[64] IARC. Diesel and gasoline engine exhausts and some nitroarenes, Vol. 105. Vol. 46. Lyon, France; 2012

[65] IARC. Some chemicals present in industrial and consumer products, food and drinking-water. Vol. 101. Lyon, France; 2013

[66] Spracklen DV, Carslaw KS, Kulmala M, Kerminen V-M, Mann GW, Sihto S-L. The contribution of boundary layer nucleation events to total particle concentrations on regional and global scales. Atmospheric Chemistry and Physics Discussions. 2006;**6**:7323-7368

[67] Curtius J. Nucleation of atmospheric aerosol particles. Comptes Rendus Physique. 2006;**7**: 1027-1045

[68] Marti JJ, Jefferson A, Cai XP, Richert C, McMurry PH, Eisele F. H_2SO_4 vapor pressure of sulfuric acid and ammonium sulfate solutions. Journal of Geophysical Research – Atmospheres. 1997;**102**(D3):3725-3735

[69] Peters K, Walters C, Moldowan M. The Biomarker Guide. Biomarkers and Isotopes in the Environment and Human History, Vol. 1. 2nd ed. Cambridge: Cambridge University Press; 2005. 451 p. ISBN: 0521781582

[70] Ho KF, Ho SSH, Lee SC, Cheng Y, Chow JC, Watson JG, et al. Emissions of gas- and particle-phase polycyclic aromatic hydrocarbons (PAHs) in the Shing Mun Tunnel, Hong Kong. Atmospheric Environment. 2009;**43**(40):6343-6351

[71] Motoyama Y, Bekki K, Chung SW, Tang N, Kameda T, Toriba A, et al. Oxidative stress more strongly induced by ortho-than para-quinoid polycyclic aromatic hydrocarbons in A549 cells. Journal of Health Science. 2009;**55**(5):845-850

[72] Agency for Toxic Substances and Disease Registry. Guidance Manual for the Assessment of Joint Toxic Action of Chemical Mixtures. U.S. Department of Health and Human Services. Atlanta, GA.; 2004

[73] Robson M, Toscano W. Risk Assessment for Environmental Health [Internet]. Jossey-Bass; 2007 [cited 2017 Jul 25]. 628p. Available from: https://books.google.com.mx/books/about/Risk_Assessment_for_Environmental_Health.html?id=s_ih18SnrvcC&redir_esc=y

[74] Tomaz S, Jaffrezo JL, Favez O, Perraudin E, Villenave E, Albinet A. Sources and atmospheric chemistry of oxy- and nitro-PAHs in the ambient air of Grenoble (France). Atmospheric Environment. 2017;**161**:144-154

Chapter 5

Atmospheric N and S Deposition Fluxes in the Metropolitan Area of Monterrey, Mexico and Its Relation with Criteria Air Pollutants and Meteorological Conditions

Rosa María Cerón Bretón,
Julia Griselda Cerón Bretón, Jonathan Kahl,
Evangelina Ramírez Lara, Atl Víctor Córdova Quiroz,
Alberto Antonio Espinosa Guzmán,
Manuel Muriel García,
Gilma Gabriela Arenas Hernández,
José Angel Solís Canul and Abril Rodríguez Guzmán

Additional information is available at the end of the chapter

http://dx.doi.org/10.5772/intechopen.79484

Abstract

The objective of this study was to assess the spatial and temporal variability of N and S atmospheric deposition and its relation with criteria air pollutants (CAPs) and meteorological conditions (MCs) in the metropolitan area of Monterrey (MAM). Atmospheric deposition was collected in 10 sampling sites during 3 climatic periods by using passive samplers based on ionic exchange resins (IERs); simultaneously, CAP and MC were monitored. Ions were extracted from IER to determine nitrate, ammonium, and sulfate levels, and deposition fluxes were estimated. On the other hand, from CAP and MC, wind and concentration roses were built to identify the exceedances of the current regulations, and relationships between CAP and meteorological conditions. It was found that only S deposition fluxes exceeded critical load values proposed in Europe, suggesting that S deposition could be a serious threat in MAM. It was found that CO in Juárez sampling site and O_3 and PM_{10} in all sampling sites showed exceedances of the current regulatory limits, showing seasonal and spatial patterns similar to N and S deposition fluxes. Deposition fluxes were mapped to identify critical zones or periods in which these fluxes could be higher as a result of the prevailing meteorological conditions.

Keywords: N deposition, S deposition, criteria pollutants, Monterrey, Mexico

1. Introduction

Atmospheric deposition of nitrogen and sulfur is a growing and significant problem for the environment in many parts of the world. However, in urban areas it has become a concern due to the increase in atmospheric emission of gases and particulate that entails consequences for the environment and for the health of population. Sulfur dioxide (SO_2), nitrogen oxide (NO_x), and ammonia (NH_3) are usually produced by anthropogenic activities. Industrial activities, vehicular emissions, and the burning of biomass are just some of the main sources of these pollutants in the atmosphere. The deposition of S and N occurs as a result of removal processes, either from precipitation (wet deposition) or from the deposit of particulate material or gas adsorption (dry deposition), and is associated with the acidification of soils and surface waters. Deposition of sulfur compounds results in the modification of the chemistry and biology of soil and water bodies such as the decrease in pH. On the other hand, the deposition of nitrogen compounds causes changes through the direct acidification of soils and natural water, or the saturation of nitrogen in vegetation species, which leads to the loss of vitality of diverse ecosystems. In addition, the deposition of N and S can cause deterioration to historical monuments and diverse materials [1]. Despite its importance, in Mexico, the monitoring of the deposition of these compounds, as well as the evaluation of their spatial and temporal distribution, and the estimation of their effects on ecosystems have not been sufficiently studied. Although in Mexican territory there are many cities with significant urban and industrial development, many of which are close to valuable historical heritage or important ecological zones, with the exception of the surrounding areas to the metropolitan zone of the Valley of Mexico, there are few air pollution studies available [2] One of the main reasons that limits the study of wet atmospheric deposition is that their study requires expensive automatic samplers that require compliance with certain specifications for installation and operation; while, in the case of dry deposition, standardized techniques are not available. In this regard, some authors [3] have proposed the use of passive samplers based on ion exchange resins for the monitoring of atmospheric deposition, this type of device allows to study several points simultaneously due to its low cost and simple design.

On the other hand, the state of Nuevo Leon has been characterized by its accelerated urban and industrial growth, which places it within the three main metropolitan areas of the country and the second with the greatest territorial extension. Additionally, the city of Monterrey is the second city in the country with the highest reports of air pollution and subsequent effects not only on public health but also on ecosystems. Previous studies in this region have shown significant correlations between the wind direction and temperature inversions and contaminant transport from regional sources. That is, the pollutants in the MAM have a seasonal component as a result of the influence of these transport processes, resulting in a greater concentration and deposition of pollutants at certain climatic periods of the year. Likewise, in addition to the contribution by regional transport, there are also significant emissions from local industrial sources and vehicular sources that may result in background levels above the reference values considered as acceptable. However, since in the case of atmospheric deposition, it is not a criterion contaminant, that is, there is not a standard or reference value that

http://dx.doi.org/10.5772/intechopen.79484

regulates it, it is necessary to carry out monitoring studies at a medium or long term to establish a baseline that allows to perform an environmental diagnosis of the area and infer its possible effects. Notwithstanding, in the MAM, some authors [4–7] have measured wet and dry atmospheric deposition; these studies have been carried out at a short term and systematic measurements that allow a proper diagnosis considering the seasonal and spatial components are not available. Therefore, the present work focuses on the mapping and study of the seasonal and spatial variability of N and S atmospheric deposition in the metropolitan area of Monterrey (MAM), Nuevo Leon, during three climatic seasons (dry, rainy, and cold fronts) using passive samplers based on ionic exchange resin at ten points distributed throughout nine municipalities of MAM.

2. Study area

The metropolitan area of Monterrey (MAM) is located to the northeast of the country in Nuevo Leon (25°42′26.53 N, 100°17′29.36 W). In 2015, it registered a total of 4,437,643 inhabitants within a surface of 6357 km^2, being the third most populated city in Mexico only after Guadalajara and Mexico City; and the second in territorial extension. Worldwide, MAM occupies the 17th place, while in Latin America, it ranks number 10. Also, it was considered by Forbes in 2010 as the fourth most intelligent city in the world, with a great capacity of sustainable growth. MAM is located 913 km from Mexico City. It is known as "The City of the Mountains" due to the orographic formations existing within and in the surroundings of the city and, because of this, MAM exhibits serious air pollution problems. MAM climate

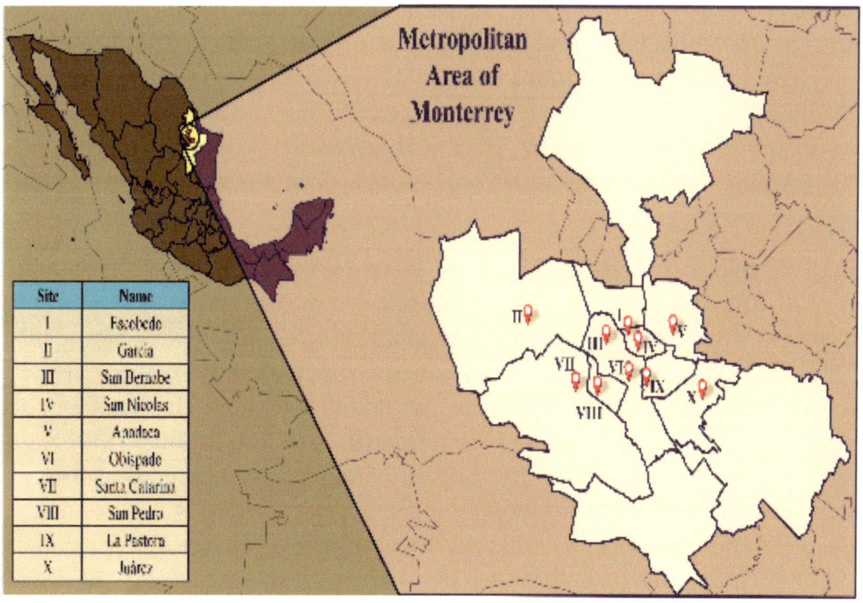

Figure 1. Sampling sites location along MAM.

is considered extreme, and according to Köppen climatic classification, it has warm and semi-arid climate (BSh), with an annual precipitation from 431.1 to 1300 mm. To assess the spatial and temporal distribution of N and S deposition fluxes, ten sampling sites were selected along MAM. The location of these sampling sites corresponds to the location of automatic monitoring stations of SIMA (Integral System of Environmental Monitoring of Monterrey). The specific location of these sampling sites and the name of each automatic monitoring station are presented in **Figure 1**.

3. Methodology

3.1. Sampling

The characterization of complex spatial patterns as atmospheric deposition of N and S in a given area requires simple monitoring equipment, which is cheap, easy to operate, and does not require frequent visits to the field. Throughfall deposition consists of solute collected in atmospheric deposition. This method is widely used to estimate the inputs of atmospheric deposit to the forests ecosystems, since, they include both, dry and wet deposition; therefore, this kind of passive sampler constitutes a good choice to obtain a reliable estimation of atmospheric inputs of N and S in a given ecosystem [8]. Passive samplers type throughfall are based in collectors of ionic exchange resin (IER). They consist of a funnel connected to a column that contains a mixed bed of ionic exchange resin (Amberlite™ IRN150). Deposition falls on the surface of the funnel, washing toward the inside of the column. The main advantage of this type of device is that it can be used during long periods of time (e.g., months) and the equipment has a very low cost, allowing to increase the number of sampling points in a given area. Therefore, with this kind of collector, it is possible to display a great number of them to characterize spatial patterns in deposition with a high resolution [9]. Nitrate, sulfate, and ammonium (NO_3^-, SO_4^{2-} y NH_4^+.) can be exchanged in IER for cations and anions, respectively, and then be trapped by functional groups with opposite electric charges. In this study, a design of a mixed resin bed was chosen, since this kind of resin captures both, anions and cations.

Throughfall deposition was collected in MAM, Nuevo Leon, Mexico, from February 26, 2017, to February 26, 2018, in ten sampling sites (**Figure 1**) which correspond to automatic monitoring stations of SIMA, by using deposition collectors based on IER operated and built according to [3, 10]. IER devices consist of funnels covered with a mesh (to prevent the fall of solid material such as leaves and insects) that are attached to PVC tubes. Inside these tubes, 30 g of IER are placed (where ions of interest are retained). Each tube is sealed with glass fiber at the bottom (as a platform or support for the resin) and at the top (as a filter). The resin tube is placed inside an outer PVC tube (shadow tube), which protects resin from solar radiation and helps avoid changes in its physical and chemical properties due to solar radiation. The lower end of the inner tube (resin tube) is closed or open by using a PVC valve to allow the hydrological flux to drain or not. Finally, resin tubes were placed in open areas at each sampling site within SIMA facilities. This exposition period allowed to obtain a data set for three seasons of 4 months each, corresponding to dry, rainy, and cold fronts or Norths seasons on an annual basis.

3.2. Chemical analysis

3.2.1. *IER extraction procedure*

To carry out this process, an extraction system specially designed for this purpose was built. This system consisted of a PVC tube 5 mm (ID) and 15 cm in length, adapted to each collector with the resin to be extracted. Glass fiber is removed with tweezers to verify that the drain hole was not dirty. After this, each resin tube is labeled and it is verified that the PVC valve is closed. All columns are placed in vertical position, and then, the resin tubes are washed with 100 ml of deionized water, allowing a repose of 20 minutes. Simultaneously, the threaded connections are revised to identify leaks. In the case of one leak identification, the joints are tightened, and if necessary, Teflon™ tape is added. Once, 20 minutes have elapsed, the valve opens so that a drip rate of 2 drops by second is obtained. A continuous drip is maintained during 10 minutes until drainage is completed. This rinse is discarded. Then, it is ensured that the PVC valve is closed, and 100 ml of 2 N KCl extraction solution is added, and allowed to repose for 20 minutes. Again, the PVC valve is open so that a drip rate of 2 drops by second is obtained. This continuous drip is maintained for 10 minutes. Finally, the valve is open to allow the remainder solution to leave the resin tube until the drainage is completed. Once, the extracts of the samples are obtained, they are stored and refrigerated at 4°C until analysis.

3.2.2. *Ammonium determination*

NH_4^+ was determined by using blue indophenol method, whose color intensity is proportional to the ammonium concentration in the sample. Determination was done by colorimetry at a wave length of 630 nm. Color formation is completed after 10 minutes and remains stable for 24 hours. Once absorbances of the samples are obtained, a quantification process was done to obtain ammonium concentrations by using a calibration curve [11].

3.2.3. *Sulfate determination*

Sulfate ion precipitates with barium chloride in an acid medium (HCl) forming crystals of barium sulfate. The spectral absorption of the barium sulfate suspension is measured at 420 nm by using a UV-Visible spectrophotometer. Sulfate concentration is determined comparing the absorbance lectures with a calibration curve, by using the turbidimetric method [12].

3.2.4. *Nitrate determination*

Nitrate anion present in the sample reacts with alkaloid brucine in an acid medium (H_2SO_4), oxidizing it and producing cocoteline, with an unstable red color, which changes quickly to yellow, being determined colorimetrically at 410 nm [13].

3.3. Meteorology analysis, criteria air pollutants and mapping

3.3.1. *Meteorology*

Speed and wind direction are determining factors in the composition of atmospheric deposition, since depending on the prevailing wind direction, it will have the influence of local continental or regional sources located upwind or the influence of maritime sources. In this study, the analysis of meteorological parameters at surface level was done by using data obtained from SIMA during the study period to identify possible anthropogenic or natural sources influencing the N and S levels found in the sampling sites. Wind roses were built to identify the prevailing wind direction in the study area. To assess the transport mechanism controlling deposition process in the study area by season, back air mass trajectories were estimated by the Lagrangian hybrid model HYSPLIT (Hybrid Single-Particle Lagrangian Integrated Trajectory) from US NOAA (National Atmospheric and Oceanic Administration).

3.3.2. *Criteria air pollutants concentration*

Database for the entire study period for each sampling point was obtained from SIMA of Monterrey for: CO, NO, NO_x, NO_2, SO_2, O_3, PM_{10} y $PM_{2.5}$. From the obtained data, concentration roses were estimated for each air pollutant and for each sampling point by climatic season to identify if daily concentrations exceeded reference values someday. These concentration roses were useful to visualize in which wind direction there were higher concentrations and, then, to identify the possible sources contributing to these levels.

3.3.3. *Statistical analysis*

A Friedman test was used to determine if atmospheric deposition fluxes were different among sampling sites, according to land use or between climatic seasons. Friedman test is a non-parametric test that can be used with block design, in which the underlying assumptions are not as restrictive as those of an ANOVA procedure (XLStat v.2017). On the other hand, principal components analysis is a technique used to reduce the dimensionality of a data set. The projection according to which data is better represented is least squares. It converts a data set of variables possibly correlated in a data set of variables without lineal correlation called principal components. Descriptive, multivariate, and principal components analysis were carried out by using XLstat-Pro v. 2017.

3.3.4. *Deposition fluxes mapping*

One of the main uses of geo-statistical mapping consists in predicting new values from variables from the sample in a given area, which is referred as spatial prediction or spatial interpolation. Spatial distribution of a variable can be modeled either using a continuous model or a discrete or mixed model. On the other hand, temporal variability makes geo-statistical mapping expensive and complex. Taking into account that the seasonal periodicity in this work is regular for the studied environmental parameters, in this case, spatial variability was analyzed for each climatic period: Dry, rainy, and cold fronts or Norths seasons. The coordinates of each sampling site and the values for N and S deposition fluxes were the inputs used to derive the specific points in the maps showing the dispersion and the measured concentration for the different studied chemical compounds. In a second step, the concentrations at neighboring sampling points within the grid were averaged to attribute a value to the point. These points were the input for the interpolation procedure [14]. The deposition contours were smoothed by using the kriging method [15]. Kriging weights were estimated from a variogram, which measures the correlation grade among sampling values in the area as a function of the distance and direction. Digital images for MAM were obtained from INEGI, and these maps were integrated to build a base map in which concentration isolines obtained from Surfer program v. 10.0 were graphed, obtaining deposition fluxes maps in each studied zone by pollutant and by climatic season.

4. Results and discussion

4.1. Sulfate deposition fluxes

4.1.1. By season

The mean S deposition flux (as sulfate) during the dry season was 27.30 ± 10.34 Kg ha^{-1} $yr.^{-1}$, with a maximum value of 47.69 Kg ha^{-1} $yr.^{-1}$ in the site labeled as VI (Obispado) at the center of MAM. The average value obtained for the rainy season was 23.65 ± 4.14 Kg ha^{-1} $yr.^{-1}$, with a maximum of 28.63 Kg ha^{-1} $yr.^{-1}$ in site I (Escobedo) located to the north of MAM. On the other hand, the mean value for S deposition flux during the Norths season was 24.15 ± 7.39 Kg ha^{-1} $yr.^{-1}$, with a maximum value of 31.48 Kg ha^{-1} $yr.^{-1}$ in the sampling site labeled as V (Apodaca), located at the northeast side of MAM. From **Figure 2a**, it was observed that S deposition fluxes showed an evident seasonality, with the highest values during the dry season, and with the lowest values along the rainy season. However, from Friedman test, since p value is major than significance levels (α = 0.05), null hypothesis (H_0) cannot be rejected; therefore, it can be concluded that there were no significant differences among S deposition fluxes by climatic season and that sulfate deposition levels have an evident influence from regional transport during all year.

4.1.2. By sampling site

In the analysis by sampling site, a mean value for S deposition flux of 25.03 ± 7.63 Kg ha^{-1} $yr.^{-1}$ was obtained. According to **Figure 2b**, it can be observed that S deposition fluxes were higher in the sampling sites labeled as VI and V, which correspond to Obispado and Apodaca at the center and northeast of MAM. By applying Friedman test, p value is major than significance level (α = 0.05), and null hypothesis cannot be rejected. Therefore, it can be concluded that there were not significant differences in S deposition fluxes among sampling sites, suggesting an evident regional influence on MAM.

4.1.3. By land use

Sampling sites were grouped depending on their land use as: Rural (sites II and VII), Urban (sites I, IV, VI, VIII, IX and X), and Industrial (sites III and V). From **Figure 2c**, it can be observed that S deposition fluxes were higher at sites with an industrial land use (sites III and V), which correspond to San Bernabé and Apodaca, located to the northwest and northeast of MAM. A Friedman test was applied, and since p value is major than

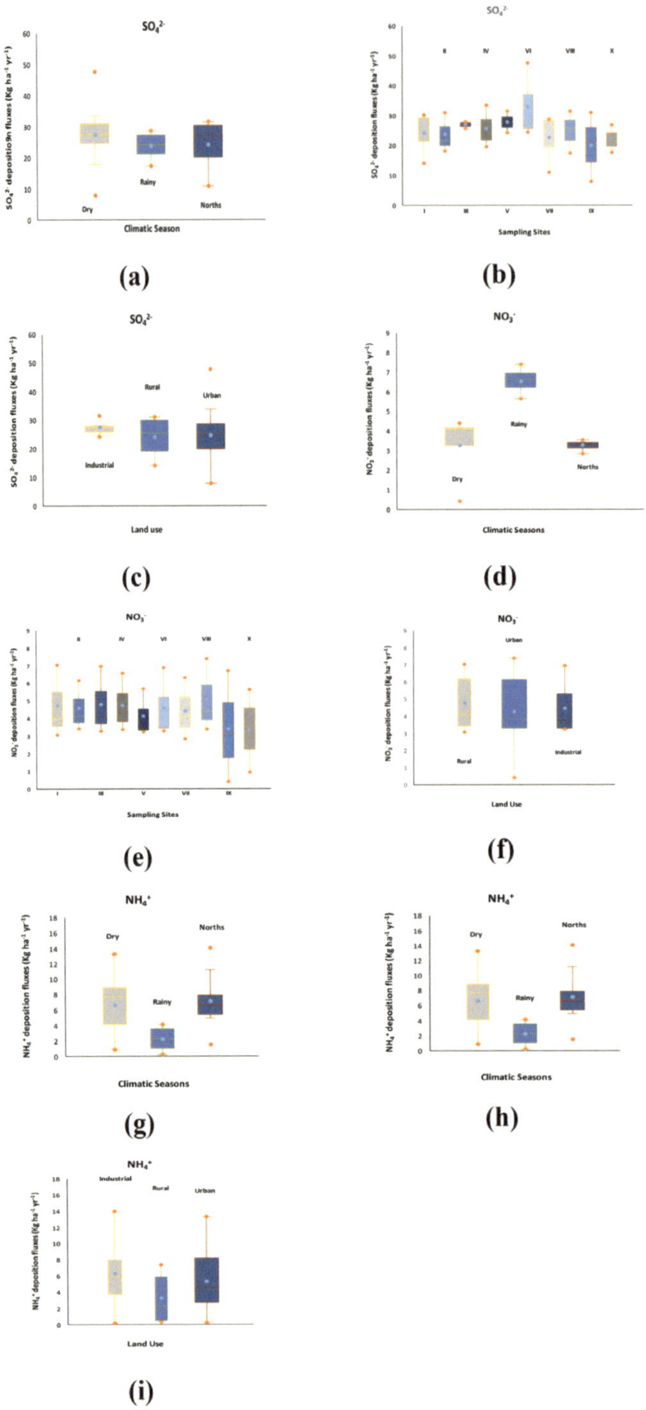

Figure 2. Sulfate deposition fluxes by: (a) climatic season, (b)sampling site, and (c) land use for MAM during the study period; Nitrate deposition fluxes by: (d) climatic season, (e) sampling site, and (f) land use for MAM during the study period; Ammonium deposition fluxes by: (g) climatic season, (h) sampling site, and (i) land use for MAM during the study period.

significance level (α = 0.05), the null hypothesis cannot be rejected, and it can be concluded that S deposition fluxes did not show significant differences by land use. This fact supports those found in the previous sections, where the regional character of sulfate due to residence time of SO_2 was completely evident.

4.2. Nitrate deposition fluxes

4.2.1. By season

The mean nitrate deposition flux value during the dry season was 3.30 ± 1.43 Kg ha^{-1} $yr.^{-1}$, with a maximum of 4.38 Kg ha^{-1} $yr.^{-1}$ corresponding to the sampling site VIII (San Pedro) located to the southwest of MAM. The average value obtained for rainy season was 6.54 ± 0.58 Kg ha^{-1} $yr.^{-1}$, with the highest value (7.39) at the sampling site VIII (San Pedro). On the other hand, during the cold fronts season, the mean value for nitrate deposition flux was 3.26 ± 0.21 Kg ha^{-1} $yr.^{-1}$, with a maximum value of 3.52 Kg ha^{-1} $yr.^{-1}$ at site X (Juárez) located to the southeast of MAM. From **Figure 2d**, it can be observed that nitrate deposition fluxes were higher along the rainy season. From Friedman test, it was found that p value is minor than significance level (α = 0.05); therefore, it can be concluded that there were significant differences in nitrate deposition fluxes between rainy season and the rest of the year (dry and cold fronts seasons). It suggests that reactions in aqueous phase can be important, at the same time.

4.2.2. By sampling site

From the analysis of results by sampling site, from **Figure 2e**, it was found that nitrate deposition fluxes were higher in the sites VIII and III: San Pedro to the southwest and San Bernabé to the northwest of MAM. By applying a Friedman test, it was found that p value was minor than significance level (α = 0.05); therefore, null hypothesis must be rejected and it is concluded that there were significant differences between sites. It means that the influence of local sources was important. It agrees with the residence time of NO_2 in the atmosphere, since it has been reported that nitrate is a local pollutant.

4.2.3. By land use

Sampling sites were grouped according to their land use as: Rural (sites II and VII), Urban (sites I, IV, VI, VIII, IX and X), and Industrial (sites III and V). From **Figure 2f**, it can be observed that nitrate deposition fluxes were higher in sampling sites with an urban land use (sites VIII and IX: San Pedro and La Pastora, located to the southwest and southeast of MAM). However, considering extreme values, these were found in sites with an industrial land use (most of the sites: IV, VI, VII, IX, and X). From **Figure 2f**, a great variability was observed, suggesting that local urban sources were mixed and emissions presented different magnitudes. It agrees with the different kinds of sources (industrial and urban) coexisting in this great metropolitan area. In spite of this, from Friedman test, it was found that p value was major than the significance level (α = 0.05); thus, the null hypothesis cannot be rejected, and therefore, it can be concluded that there were not significant differences between sampling sites by land use.

4.3. Ammonium deposition fluxes

4.3.1. *By season*

The mean ammonium deposition flux during the dry season was 6.90 ± 3.88 Kg ha^{-1} yr.$^{-1}$, with a maximum value of 13.31 Kg ha^{-1} yr.$^{-1}$ in the sampling site labeled as VII (Santa Catarina) located to the southwest of MAM. During the rainy season, the average of ammonium deposition flux was 2.21 ± 1.49 Kg ha^{-1} yr.$^{-1}$, with a maximum of 4.08 Kg ha^{-1} yr.$^{-1}$ in site X (Juárez) located to the southeast of MAM. Finally, during the cold fronts season, ammonium deposition fluxes presented a mean value of 7.14 ± 3.49 Kg ha^{-1} yr.$^{-1}$, with a peak value of 14.04 Kg ha^{-1} yr.$^{-1}$ in the sampling site III (San Bernabé) at the northwest side of MAM. From **Figure 2g**, it is observed that, ammonium deposition fluxes were higher during the dry and cold fronts seasons. Ammonium levels were significantly lower during the rainy season, suggesting a washing effect during this season. From Friedman test, it was found that p value was minor than the significance level ($\alpha = 0.05$), and the null hypothesis can be rejected; therefore, it is possible to conclude that there were significant differences between dry and cold fronts seasons and the rainy season. It suggests that, during the rainy season, a dilution effect could influence the ammonium deposition fluxes, considering that during the rest of the year, rains are scarce in MAM.

4.3.2. *By sampling site*

From **Figure 2h**, analyzing ammonium deposition fluxes by sampling site, the highest value was found in the sites VII and IV (Santa Catarina and San Nicolás), located to the southwest and northeast of MAM. Applying a Friedman test, it was found that p value is major than significance level ($\alpha = 0.05$), and therefore, the null hypothesis cannot be rejected. Then, it can be concluded that there were not significant differences among sampling sites.

4.3.3. *By land use*

Sampling sites were grouped according to their land use as: Rural (sites II and VII), Urban (sites I, IV, VI, VIII, IX, and X), and Industrial (sites III and V). From **Figure 2i**, it can be observed that ammonium deposition fluxes were higher in sampling sites with an industrial and urban land use. The emission of amines and NH_3 has been reported from vehicles (with the presence of a catalytic convertor that has enough stored hydrogen), where NO is reduced to NH_3, and deposited as NH_4^+. Therefore, vehicular emissions could have an important influence on ammonium deposition in MAM. According to the Friedman test, p value was major than the significance level, thus null hypothesis cannot be rejected, and therefore, it can be concluded that there were not significant differences among sampling sites considering their land use.

4.4. Meteorology and criteria air pollutants

4.4.1. *Site I Escobedo*

O_3 showed a strong seasonal variation (**Figure 3a**) with the highest values during the dry season, (0.074–0.095 ppm) exceeding the reference value for 8 hours (70 ppb) [16]. Both, O_3 and PM_{10} showed highest values when wind direction came from the East. In the case of PM_{10},

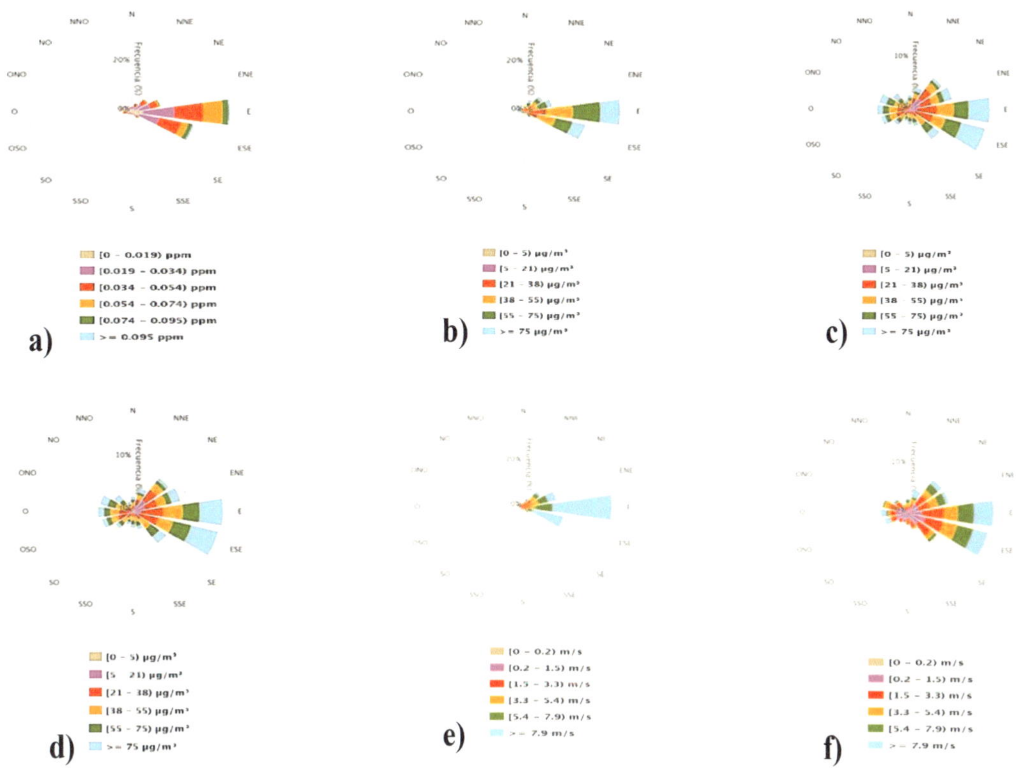

Figure 3. Criteria air pollutants and meteorological conditions for site I (Escobedo) during the study period: (a) O_3 dry season, (b) PM_{10} dry, and rainy seasons, (c) PM_{10} cold fronts season, (d) $PM_{2.5}$ cold fronts season, (e) wind dry and rainy season, (f) wind cold fronts season.

Figure 3b shows that PM_{10} levels (≥75 µg m^{-3}) exceeded the reference value for 24 hours (75 µg m^{-3}) [17]. In **Figure 3c**, it can be observed that PM_{10} levels during the cold fronts season exceeded the reference value but winds also showed a great variability. In the case of $PM_{2.5}$ (**Figure 3d**), its levels (≥75 µg m^{-3}) exceeded the reference value (45 µg m^{-3}) [17] for 24 hours during the cold fronts season when winds showed a great variability. Finally, winds came from the East during dry and rainy seasons (**Figure 3e**), showing a great variability during the cold fronts season (**Figure 3f**) with maximum wind speeds >7.9 m s^{-1} along the year.

4.4.2. Site II Garcia

From **Figure 4a** and **b**, it can be observed that O_3 levels (0.074–0.095 ppm) were higher during the dry and rainy seasons, exceeding the reference value for 8 hours (70 ppb) [16] when wind came from the East. PM_{10} levels (≥75 µg m^{-3}) were high during all year, exceeding reference value for 24 hours (75 µg m^{-3}) [17], mainly when winds came from Northeast (**Figure 4c** and **d**). Finally, winds came from the East-Northeast during dry and rainy seasons (**Figure 4e**), showing a great variability during the cold fronts season (**Figure 4f**) with maximum wind speeds >7.9 m s^{-1} along the year. Both O_3 and PM_{10} showed highest levels when winds had an East-Northeast component.

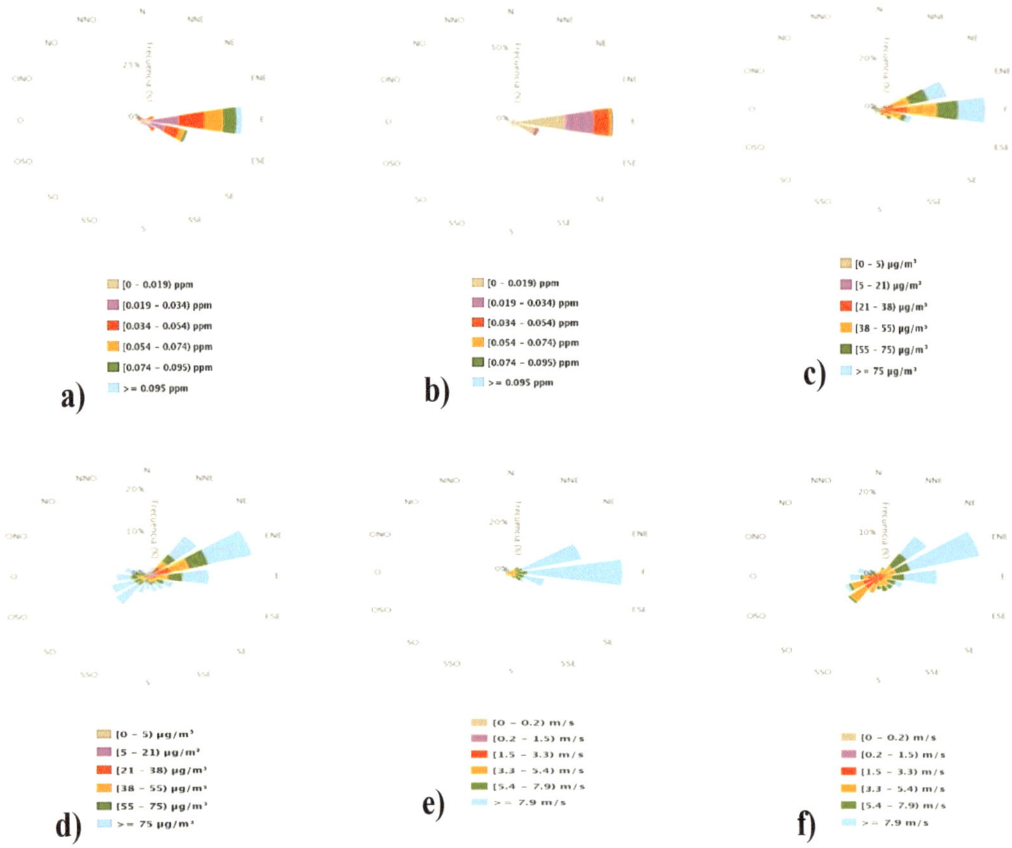

Figure 4. Criteria air pollutants and meteorological conditions for site II (García) during the study period: (a) O_3 dry season, (b) O_3 rainy season, (c) PM_{10} dry and rainy seasons, (d) PM_{10} cold fronts season, (e) wind dry and rainy season, (f) wind cold fronts season.

4.4.3. Site III San Bernabe

O_3 levels (0.074–0.095 ppm) were higher during all year, exceeding the reference value for 8 hours (70 ppb) [16] when winds came from East-Southeast (**Figure 5a** and **b**). In addition, PM_{10} levels (≥75 μg m^{-3}) also exceeded the reference value for 24 hours (75 μg m^{-3}) [17] during all year, but showing highest values and a great variability in wind direction during the cold fronts season (**Figure 5c** and **d**). Finally, winds came from the East-Southeast during dry and rainy seasons (**Figure 5e**), showing a great variability during the cold fronts season (**Figure 5f**) with maximum wind speeds >7.9 m s^{-1} along the year. Both O_3 and PM_{10} showed highest levels when winds had an East-Southeast component.

4.4.4. Site IV San Nicolas

O_3 levels (0.074–0.095 ppm) were high during all year, exceeding the reference value for 8 hours (70 ppb) [16] when winds came from North and East (**Figure 6a–c**). In addition, PM_{10} levels

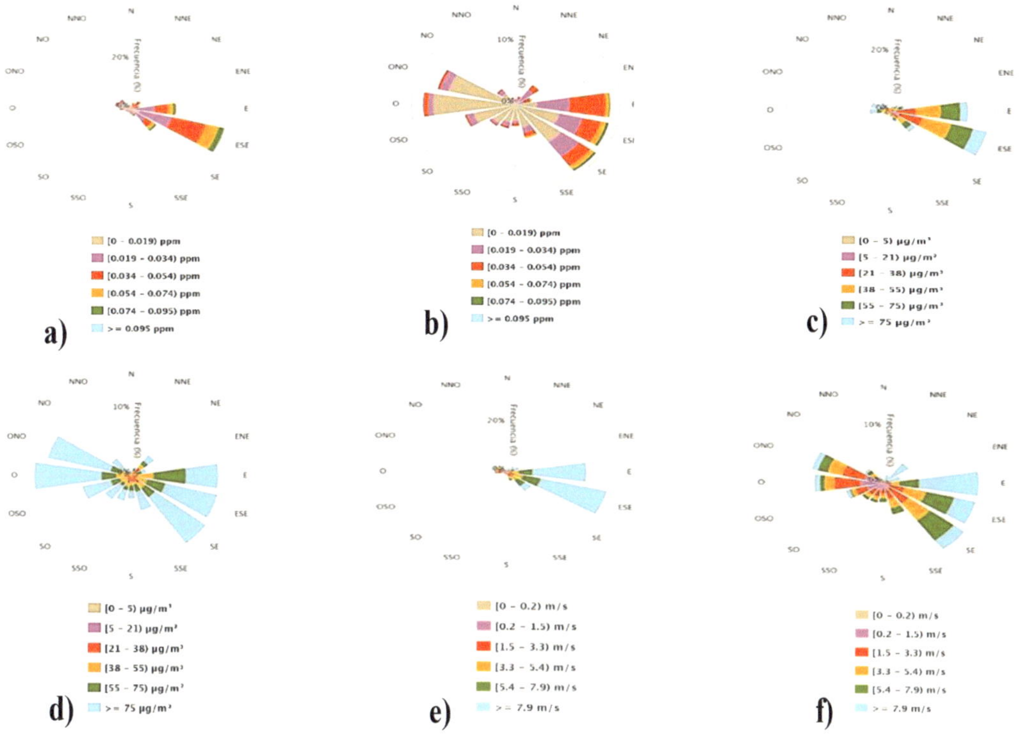

Figure 5. Criteria air pollutants and meteorological conditions for site III (San Bernabé) during the study period: (a) O_3 dry and rainy season, (b) O_3 cold fronts season, (c) PM_{10} dry and rainy seasons, (d) PM_{10} cold fronts season, (e) wind dry and rainy seasons, (f) wind cold fronts season.

(≥75 µg m^{-3}) also exceeded the reference value for 24 hours (75 µg m^{-3}) [17] during all year when winds came from East and North, but showing highest values and a great variability in wind direction during the cold fronts season (**Figure 6d–f**). Finally, winds came from the North during dry season and from East during the rainy season (**Figure 6g** and **h**), showing a great variability during the cold fronts season (**Figure 6i**) with maximum wind speeds >7.9 m s^{-1} along the year. Both O_3 and PM_{10} showed highest levels when winds had a North and East component.

4.4.5. Site V Apodaca

From **Figure 7a**, it can be observed that O_3 levels (≥0.095 ppm) were higher during the dry season when winds came from Northeast, exceeding the reference value for 8 hours (70 ppb) [16]. PM_{10} levels (≥75 µg m^{-3}) were high during all year, exceeding reference value for 24 hours (75 µg m^{-3}) [17]; mainly when winds came from North and Northwest (**Figure 7b** and **c**). $PM_{2.5}$ levels (≥44 µg m^{-3}) were high during rainy and cold fronts seasons when wind direction was from North, exceeding the reference value (45 µg m^{-3}) [17] for 24 hours (**Figure 7d**). Winds came from the Northwest and North during dry and wet (rainy and cold fronts) seasons, respectively (**Figure 7e** and **f**), with maximum wind speeds >7.9 m s^{-1} along these seasons. O_3, $PM_{10,}$ and $PM_{2.5}$ showed highest levels when winds had a North-Northwest component.

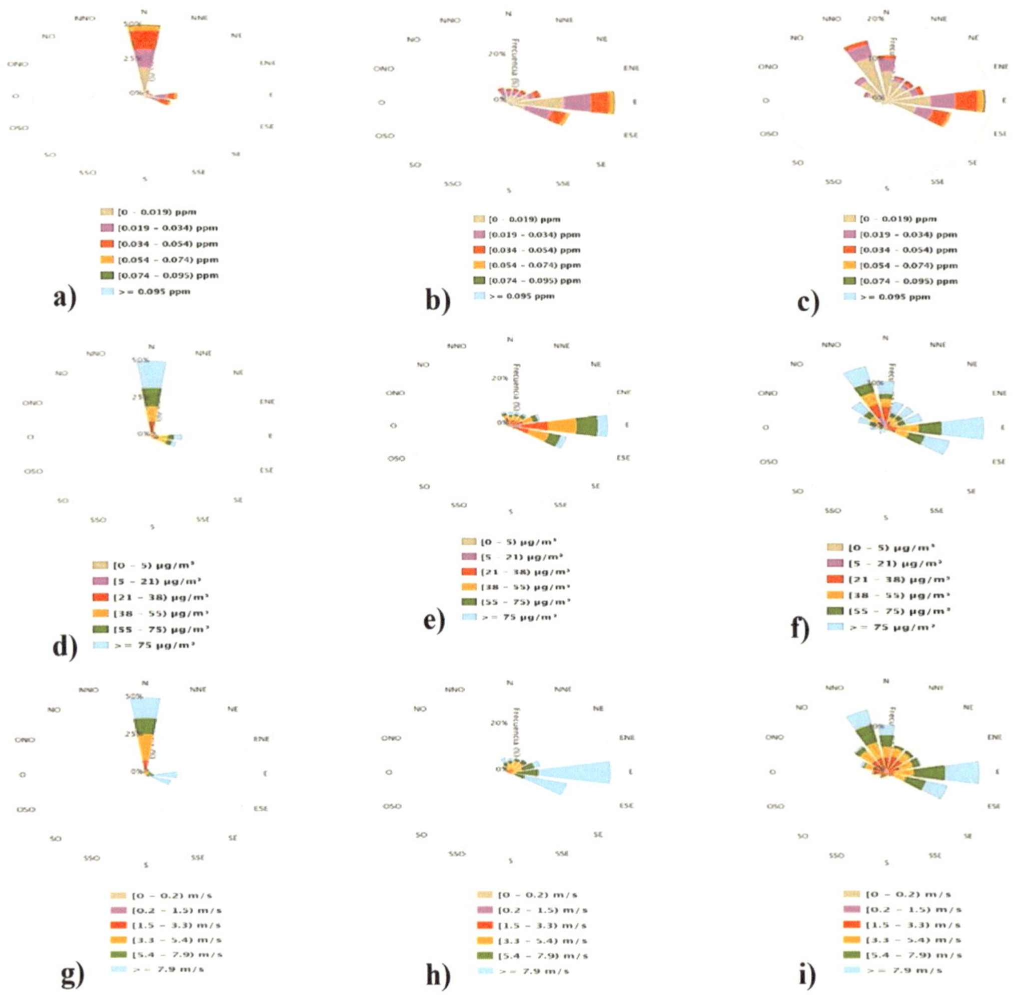

Figure 6. Criteria air pollutants and meteorological conditions for site IV (San Nicolás) during the study period: (a) O_3 dry season, (b) O_3 rainy season, (c) O_3 cold fronts season, (d) PM_{10} dry season, (e) PM_{10} rainy season, (f) PM_{10} cold fronts season, (g) wind dry season, (h) wind rainy season, (i) wind cold fronts season.

4.4.6. Site VI Obispado

From **Figure 8a**, it can be observed that O_3 levels ($\geq$0.095 ppm) were higher during the dry season when winds came from Northeast, exceeding the reference value for 8 hours (70 ppb) [16]. PM_{10} levels ($\geq$75 µg m^{-3}) were high during all year, exceeding reference value for 24 hours (75 µg m^{-3}) [17], mainly when winds came from Northeast and Southwest (**Figure 8b**). $PM_{2.5}$ levels were $\geq$44 µg m^{-3} during all year when wind direction was from Northeast, exceeding the reference value (45 µg m^{-3}) [17] for 24 hours (**Figure 8c** and **d**). In addition, winds came from the Northeast during all year, showing a great variability during cold fronts season (**Figure 8e** and **f**), with maximum wind speeds >7.9 m s^{-1} along this season. O_3, PM_{10}, and $PM_{2.5}$ showed highest levels when winds had a Northeast component.

http://dx.doi.org/10.5772/intechopen.79484

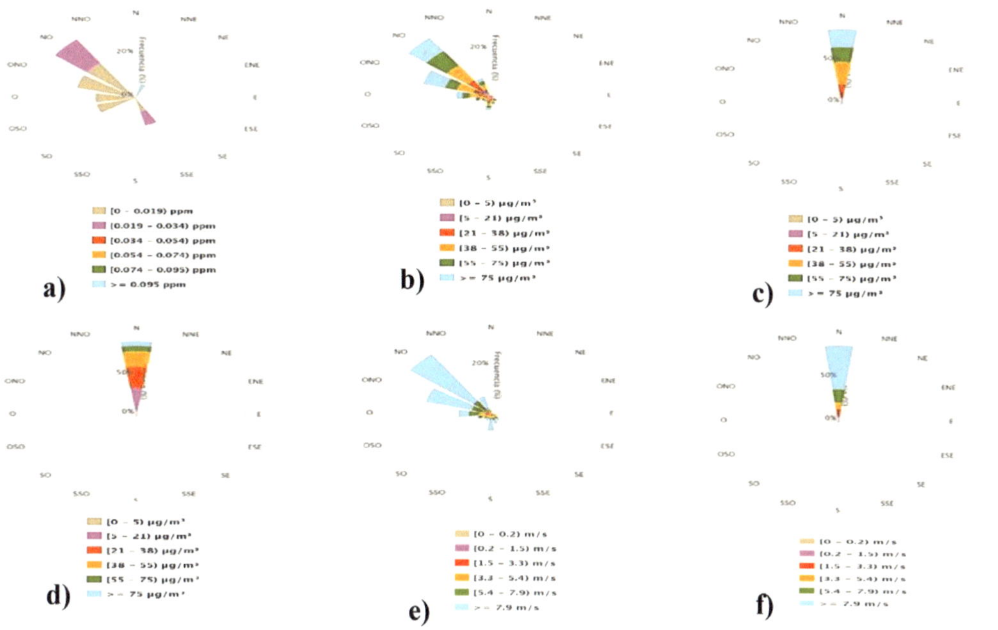

Figure 7. Criteria air pollutants and meteorological conditions for site V (Apodaca) during the study period: (a) O_3 dry season, (b) PM_{10} dry season, (c) PM_{10} rainy and cold fronts seasons, (d) $PM_{2.5}$ rainy and cold fronts seasons, (e) wind dry season, (f) wind rainy and cold fronts seasons.

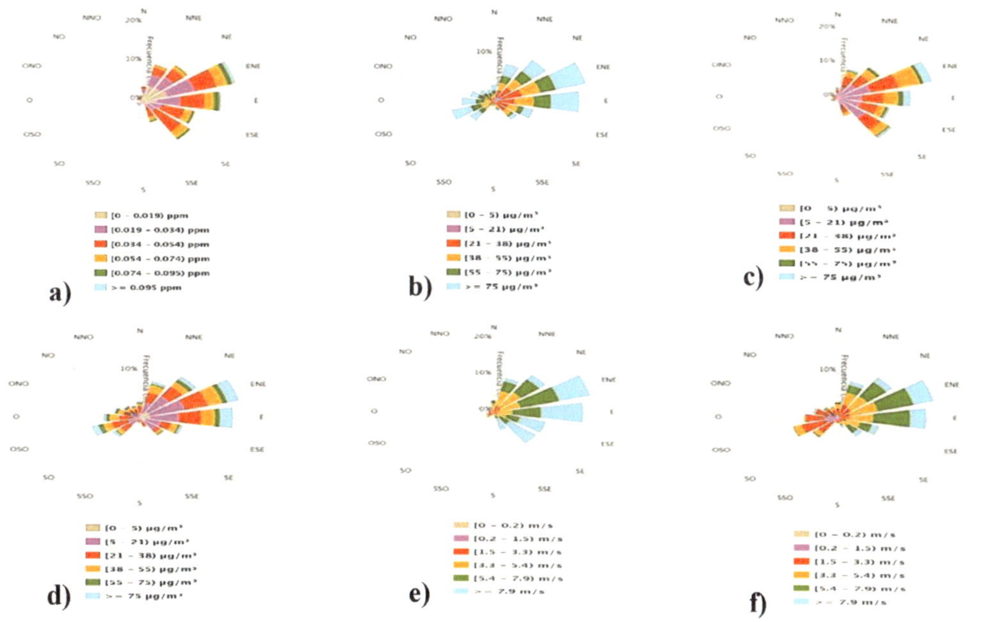

Figure 8. Criteria air pollutants and meteorological conditions for site VI (Obispado) during the study period: (a) O_3 dry, rainy and cold fronts seasons, (b) PM_{10} dry, rainy and cold fronts seasons, (c) $PM_{2.5}$ dry and rainy seasons, (d) $PM_{2.5}$ cold fronts season, (e) wind dry and rainy season, (f) wind cold fronts season.

4.4.7. Site VII Santa Catarina

O_3 levels (≥0.095 ppm) were high during all year, being higher during the dry and rainy seasons, exceeding the reference value for 8 hours (70 ppb) [16] when winds came from North (**Figure 9a** and **b**). In addition, PM_{10} levels (≥75 μg m^{-3}) also exceeded the reference value for 24 hours (75 μg m^{-3}) [17] during all year and mainly when winds came from North (**Figure 9c**). $PM_{2.5}$ levels were ≥44 μg m^{-3} during dry season when wind direction was from North, exceeding the reference value (45 μg m^{-3}) [17] for 24 hours (**Figure 9d**). According to **Figure 9e**, O_3 levels decreased significantly during the rainy season without showing exceedances to reference value. Finally, winds came from the North during all year (**Figure 9f**), with maximum wind speeds >7.9 m s^{-1} along the year. O_3, PM_{10}, and $PM_{2.5}$ showed highest levels when winds had a North component.

4.4.8. Site VIII San Pedro

From **Figure 10a** and **b**, it can be observed that O_3 levels (0.074–0.095 ppm) were higher during the dry season when winds came from East-Northeast, exceeding the reference value for 8 hours (70 ppb) [16]. PM_{10} levels (≥75 μg m^{-3}) were high during all year, exceeding reference value for 24 hours (75 μg m^{-3}) [17]; mainly when winds came from East-Northeast (**Figure 10c** and **d**). Finally, winds came from the Northeast during all year, showing a great variability during cold fronts season (**Figure 10e** and **f**), with maximum wind speeds >7.9 m s^{-1} along this season. Both O_3 and PM_{10} showed highest levels when winds had a Northeast component.

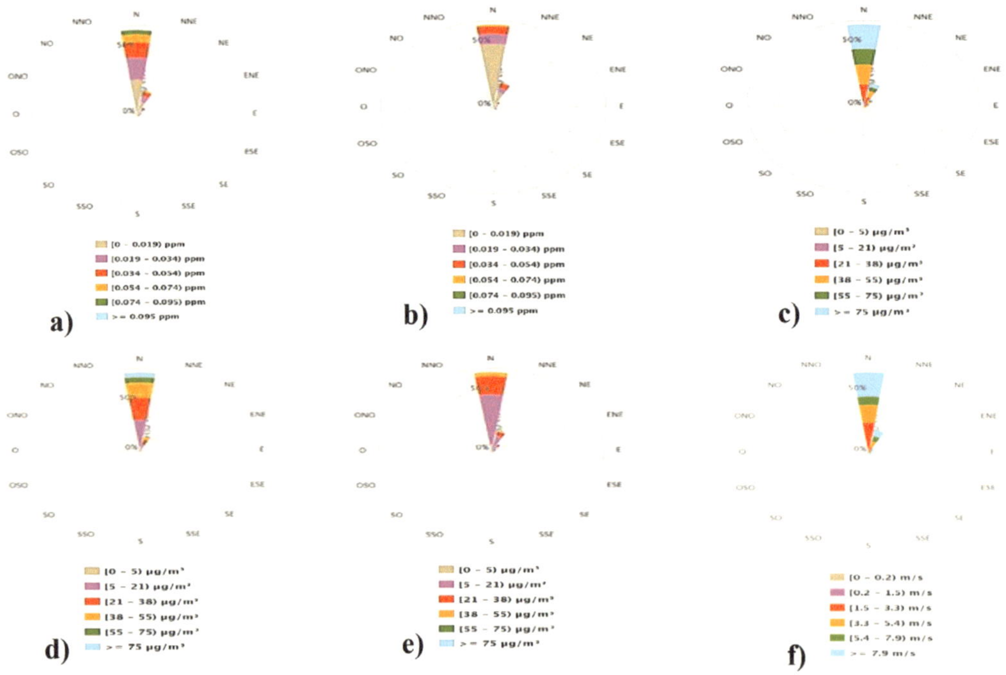

Figure 9. Criteria air pollutants and meteorological conditions for Site VII (Santa Catarina) during the study period: (a) O_3 dry and rainy seasons, (b) O_3 cold fronts season, (c) PM_{10} dry, rainy and cold fronts seasons, (d) $PM_{2.5}$ dry season, (e) $PM_{2.5}$ rainy season, (f) wind all year.

http://dx.doi.org/10.5772/intechopen.79484

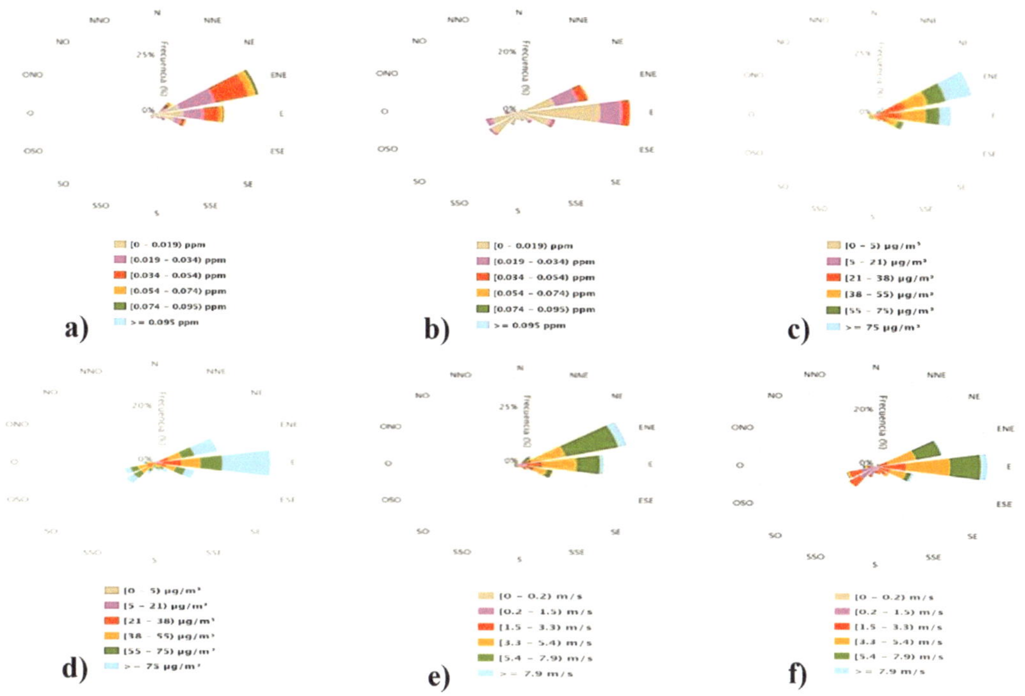

Figure 10. Criteria air pollutants and meteorological conditions for Site VIII (San Pedro) during the study period: (a) O_3 dry and rainy seasons, (b) O_3 cold fronts season, (c) PM_{10} dry and rainy seasons, (d) PM_{10} cold fronts season, (e) wind dry and rainy seasons, (f) wind cold fronts season.

4.4.9. Site IX La Pastora

O_3 levels (0.074–0.095 ppm) were high during dry and rainy seasons, exceeding the reference value for 8 hours (70 ppb) [16] when winds came from North-Northeast (**Figure 11a**). In addition, PM_{10} levels (≥75 µg m^{-3}) were high during all year and also exceeded the reference value for 24 hours (75 µg m^{-3}) [17], when winds came from North-Northeast (**Figure 11b**). $PM_{2.5}$ levels were ≥44 µg m^{-3} during all year, being higher during cold fronts season, and when wind direction was from North-Northeast, exceeding the reference value (45 µg m^{-3}) [17] for 24 hours (**Figure 11c**). Finally, winds came from the North-Northeast during all year (**Figure 11d–f**) with maximum wind speeds >7.9 m s^{-1} along the year, and showing a greater variability in wind direction during cold fronts season. O_3, PM_{10}, and $PM_{2.5}$ showed highest levels when winds had a Northeast component.

4.4.10. Site X Juarez

Figure 12a shows that CO levels (8.5–11 ppm) were higher during cold fronts season, reaching the upper limit value established in the air quality standard (11 ppm) for 8 hours [18]. From **Figure 12b** and **c**, it can be observed that O_3 levels (≥0.095 ppm) were high during dry season when winds came from Southeast, exceeding in both cases, the reference value for 8 hours (70 ppb) [16]. PM_{10} levels (≥75 µg m^{-3}) were high during all year, exceeding

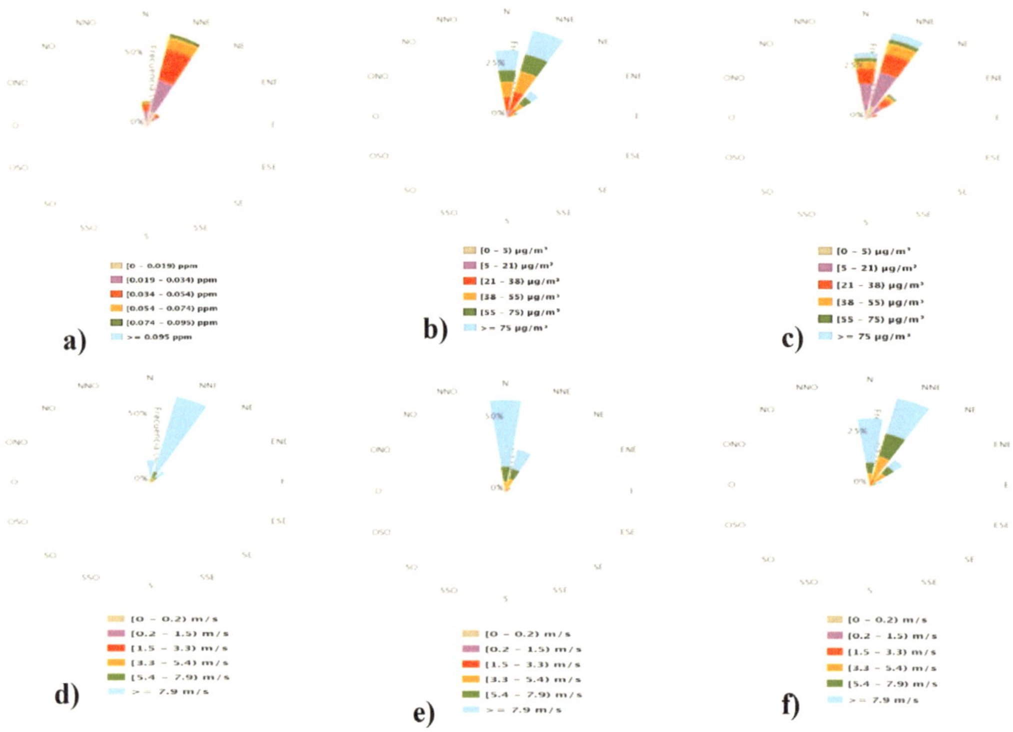

Figure 11. Criteria air pollutants and meteorological conditions for site IX (La Pastora) during the study period: (a) O_3 dry, rainy and cold fronts seasons, (b) PM_{10} dry, rainy, and cold fronts seasons, (c) $PM_{2.5}$ dry, rainy, and cold fronts seasons, (d) wind dry season, (e) wind rainy season, (f) wind cold fronts season.

reference value for 24 hours (75 µg m^{-3}) [17], when winds came from Southeast during dry and rainy season (**Figure 12d**) and from Northwest and Southeast during cold fronts season, showing a greater variability in wind direction (**Figure 12e**). Finally, winds came from the Southeast during dry and rainy seasons, and from Northwest during cold fronts season (**Figure 12f**), with maximum wind speeds >7.9 m s^{-1} along the year. O_3 and PM_{10} showed highest levels when winds had a Southeast component most part of the year and a Northwest component during cold fronts season, suggesting a seasonal behavior for these pollutants. However, in the case of CO behavior, it was completely different, with the highest levels (even exceeding the air quality standard) during cold fronts season with winds coming from Southeast and Northwest.

4.5. Mapping N and S deposition fluxes and reference values

In Mexico, reference values to compare the current deposition fluxes of N and S are not available. However, critical loads have been estimated for European ecosystems and some sites in the United States. A critical load value of 5 Kg N ha^{-1} $yr.^{-1}$ has been proposed for alpine ecosystems [19], whereas for some sites in North America, values of 3–8 Kg N ha^{-1} $yr.^{-1}$ for New Mexico and 4–7 Kg N ha^{-1} $yr.^{-1}$ for California have been proposed [20].

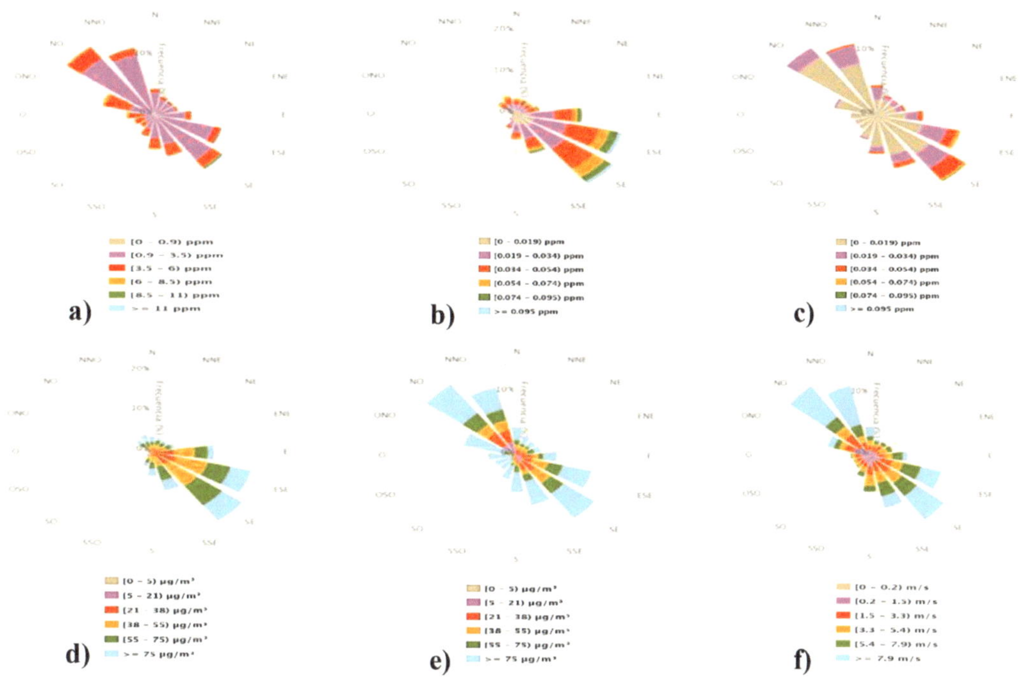

Figure 12. Criteria air pollutants and meteorological conditions for site X (Juarez) during the study period: (a) CO cold fronts season, (b) O_3 dry season, (c) O_3 rainy and cold fronts seasons, (d) PM_{10} dry and rainy seasons, (e) PM_{10} cold fronts season, (f) wind all year.

In the case of S deposition, a critical value of 3 Kg S ha^{-1} $yr.^{-1}$ has been reported for very sensitive areas in Europe, whereas for natural forests, a reference value of 2–5 Kg S ha^{-1} $yr.^{-1}$ has been proposed [21]. In this study, mean N and S throughfall deposition fluxes were 4.88 and 25.03 Kg ha^{-1} $yr.^{-1}$, respectively. N deposition fluxes did not exceed the reference value reported for alpine ecosystems; however, they are almost in the upper limit of this reference value and similar to those found in New Mexico and California. In addition, N deposition levels found in MAM (**Figure 13**) are almost twice those reported by Escoffie [22] in Carmen Island (2.15 Kg N ha^{-1} $yr.^{-1}$), Campeche; by Sánchez [23] in Orizaba Valley, Veracruz (1.44 Kg N ha^{-1} $yr.^{-1}$); and by López [24] in Mérida, Yucatán (2.7 Kg N ha^{-1} $yr.^{-1}$) and are almost four times those reported by García [25] in Atasta-Xicalango, Campeche (1.15 Kg N ha^{-1} $yr.^{-1}$). On the other hand, S deposition fluxes in MAM exceeded almost eight times the critical load proposed for sensitive areas, and five times the upper reference value for natural forests in Europe. S deposition fluxes found in MAM were almost six times higher than those reported by Escoffie [22] in Carmen Island, Campeche (4.7 Kg S ha^{-1} $yr.^{-1}$); and by López [24] in Mérida, Yucatán (4.07 Kg S ha^{-1} $yr.^{-1}$), and almost three times higher than those reported by García [25] in Atasta-Xicalango, Campeche (8.57 Kg S ha^{-1} $yr.^{-1}$). In spite of S levels in MAM being half of those reported by Sánchez [23] in Orizaba Valley (55.16 Kg S ha^{-1} $yr.^{-1}$), the current S deposition fluxes in MAM represent a risk potential of acidification and impact on ecosystems in this region.

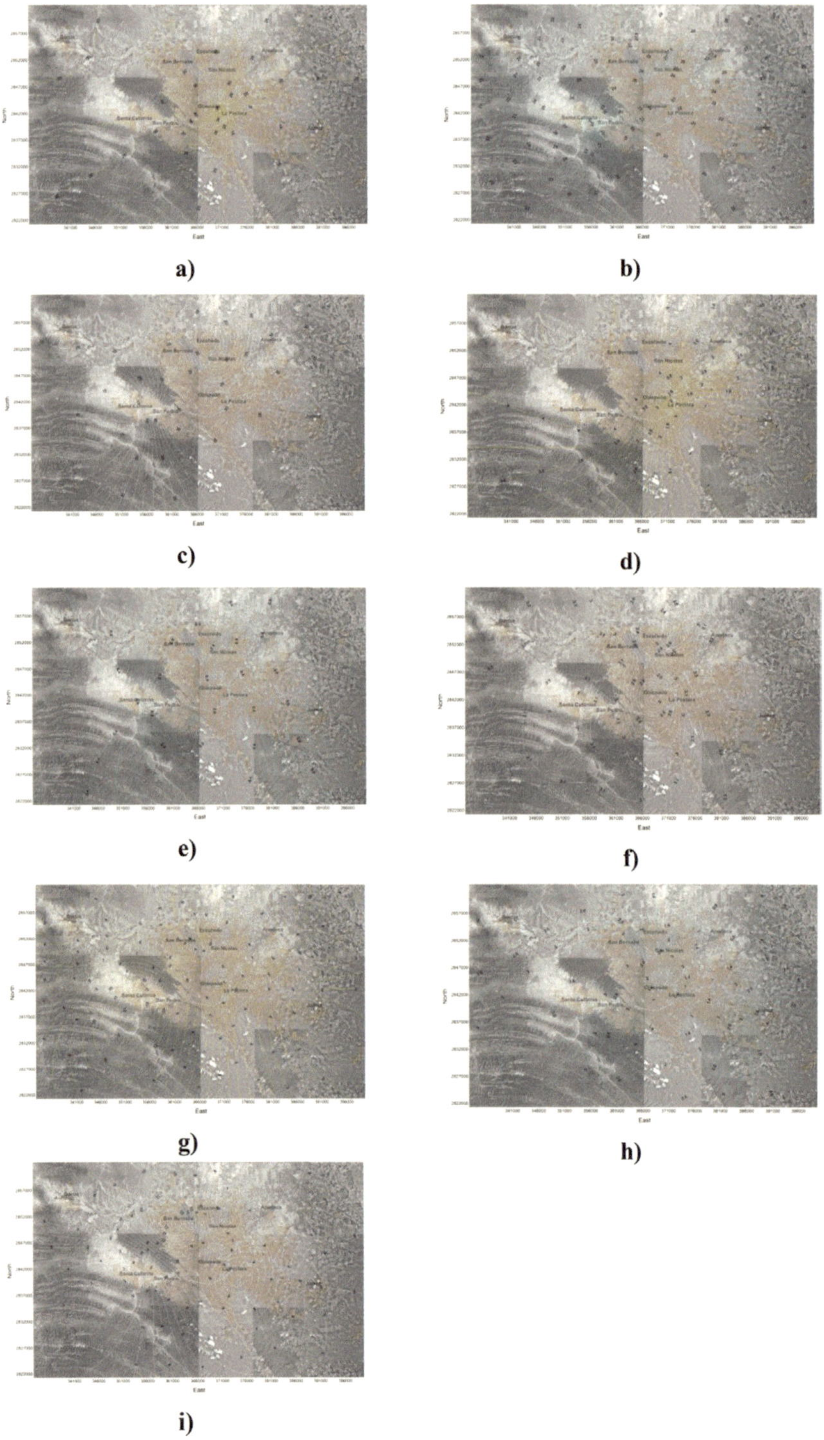

Figure 13. Spatial and temporal patterns for throughfall deposition fluxes of SO_4^{2-} for (a) dry season, (b) rainy season, (c) cold fronts season; of NO_3^- for (d) dry season, (e) rainy season, (f) cold fronts season; and of NH_4^+ for (g) dry season, (h) rainy season, (i) cold fronts season in MAM during the study period.

http://dx.doi.org/10.5772/intechopen.79484

5. Conclusions

This chapter presents an overview of atmospheric pollution and its spatial and temporal variability in MAM, and from results, we can conclude that:

N deposition fluxes: Nitrate deposition showed a seasonal pattern with the highest levels during the rainy season (suggesting that atmospheric reactions in aqueous phase play an important role in the removal process). In the case of ammonium, its deposition also presented a seasonal variation, with higher levels during the dry and cold fronts season in Santa Catarina municipality. N deposition fluxes did not exceed the critical load values reported for Europe and USA; however, these levels were higher than those reported for the southeast region of Mexico.

S deposition fluxes: Sulfate deposition did not show significant differences between seasons and sampling points, suggesting that levels found probably correspond to background levels in MAM. Sulfate levels were relatively high in Obispado, Santa Catarina, and Escobedo municipalities. S deposition fluxes exceeded the limit values proposed for sensitive areas and natural forests in Europe, and were higher than those reported at the southeast (SE) of the country, but lower than those found at the center of Mexico. It suggests that S deposition could be a potential risk for ecosystems and historical heritage in MAM.

CO: Juárez municipality was the only sampling site that showed exceedances to the reference value established in the current regulation, this municipality is located to the east of MAM, and its levels were higher when wind came from N.

O_3: Ozone levels exceeded the reference value of the current regulation in all sampling sites during the dry season when wind had an east component (E-SE-NE).

PM_{10}: PM_{10} levels exceeded the threshold value of the current regulation in all sites and during all year, its levels being higher when wind came from East (E-SE-NE).

$PM_{2.5}$: Obispado and La Pastora municipalities (center of MAM) showed the highest levels during all year, whereas in Escobedo and Apodaca (at the northern side of MAM), its levels were higher during the cold fronts season.

In spite of the time scale in which deposition fluxes (by season) and criteria pollutants (by day) were different, we could identify an evident association between CO and nitrate, since both analysis showed that their levels were higher in Juarez municipality during cold fronts season (CO levels exceeded the regulation's reference values and exhibited a different pattern regarding to the remaining sampling sites in MAM). It suggests that both, CO and nitrate had their origin in vehicular sources in this urbane zone highly polluted. On the other hand, a similarity was observed between deposition patterns of S and PM_{10}-$PM_{2.5}$ levels in MAM, since sulfate did not present significant differences in its spatial and seasonal variability; it suggests that levels found in this study remained constant all year, and correspond to the background levels for MAM. The same finding was obtained for PM_{10} and $PM_{2.5}$ levels, since their levels exceeded the reference value established in the current regulation in all sampling sites. Regarding wind direction, an evident association with criteria pollutants was found, PM_{10} and O_3 showed their highest levels when wind had an east component (E-SE-NE), which corresponds to the prevailing wind direction during all year in MAM. In addition, $PM_{2.5}$ levels were higher when wind came from north. It suggests that sources located at north (N) and east (E) from MAM contributed significantly to pollution in MAM. Finally, this study suggests that, since O_3 and PM_{10} levels exceeded the allowable maximum limit during all year and in all sampling sites, the implications that this fact may have on the population health in MAM could be serious.

Author details

Rosa María Cerón Bretón[1*], Julia Griselda Cerón Bretón[1], Jonathan Kahl[2], Evangelina Ramírez Lara[3], Atl Víctor Córdova Quiroz[1], Alberto Antonio Espinosa Guzmán[4], Manuel Muriel García[5], Gilma Gabriela Arenas Hernández[1], José Angel Solís Canul[1] and Abril Rodríguez Guzmán[1]

*Address all correspondence to: rosabreton1970@gmail.com

1 Autonomous University of Carmen, Ciudad del Carmen, Campeche, Mexico

2 University of Wisconsin-Milwaukee, Milwaukee, Wisconsin, USA

3 Automous University of Nuevo Leon, Monterrey, Nuevo Leon, Mexico

4 Research Center on Corrosion, Campeche, Mexico

5 Mexican Institute of Petroleum, Ciudad del Carmen, Campeche, Mexico

References

[1] Cerón RM, Cerón JG, Muriel M, Anguebes F, Ramírez M, Zavala J, et al. Spatial and temporal distribution of throughfall deposition of nitrogen and sulfur in the mangrove forests associated to Terminos Lagoon. In: Nejadkoorki F, editor. Current Air Quality Issues. 1st ed. Rijeka Croacia: IntechOpen; 2015. pp. 147-164. DOI: 10.5772/59726

[2] Fenn M, Bauer L, Quevedo A, Rodriguez C. Nitrogen and sulfur deposition and forest nutrient status in the Valley of Mexico. Water, Air, and Soil Pollution. 1999;**113**:155-174. DOI: 10.1023/A:1005033008277

[3] Fenn M, Poth M. Monitoring nitrogen deposition in throughfall using ion exchange resin columns: A field test in the San Bernardino Mountains. Journal of Environmental Quality. 2004;**33**:2007-2014. DOI: 10.2134/jeq2004.2007

[4] Ramírez E, Miranda R, Gracia Y, Balderas I, Bravo H, Sosa R, et al. Chemical composition of rainwater in northeastern Mexico. Revista Atmósfera. 2010;**23**:213-224

[5] Cerón R, Cerón J, Aguilar C, Ramírez E, Ortínez J, Montalvo C, et al. Wet deposition fluxes and related atmospheric chemistry at three sites in Mexico. Open Journal of Air Pollution. 2014a;**3**:1-9. DOI: 10.4236/ojap.2014.31001

[6] Cerón R, Cerón J, Ramírez E, Aguilar C, Montalvo C, López U, et al. Variations in criteria pollutants and deposition fluxes of trace elements in metropolitan area of Monterrey, Mexico. In: Proceedings of the 7th International Conference on Environmental and Geological Science and Engineering (EG'14); 3-5 June 2014; Salerno, Italy. Salerno: WSEAS; 2014. pp. 321-328

[7] Casanova M, Corroy C. Análisis de la Depositación Atmosférica Seca en el Municipio de San Nicolás de los Garza, Nuevo León [thesis]. Ciudad del Carmen Campeche: Universidad Autónoma del Carmen; 2017

[8] Butler T, Likens G. A direct comparison of throughfall plus stemflow to estimates of dry and total deposition for sulfur and nitrogen. Atmospheric Environment. 1995;**29**:1253-1265

[9] Clow D, Roop H, Nanus L, Fenn M, Sextone G. Spatial patterns of atmospheric deposition of nitrogen and sulfur using ion-exchange resin collectors in Rocky Mountain National Park, USA. Atmospheric Environment. 2015;**101**:149-157. DOI: 10.1016/j.atmosenv.2014.11.027

[10] Simkin S, Lewis D, Weathers K, Lovett G, Schwartz K. Determination of sulfate, nitrate and chloride in throughfall using ion-exchange resins. Water, Air, and Soil Pollution. 2004;**153**:343-354. DOI: 10.1023/B:WATE.0000019958.59277

[11] Fresenius W, Quentin K, Schneider W. Water Analysis; A Practical Guide to Physico-chemical, Chemical and Microbiological Water Examination and Quality Assurance. 1st ed. Berlin/Heidelberg: Springer-Verlag; 1988. pp. 195-476. DOI: 10.1007/978-3-642-72610-1

[12] NMX-AA-074-SCFI-2014. Medición del ión sulfato en aguas naturales, residuales y residuales tratadas–Método de prueba. Dirección General de Normas. México, D.F; 2015

[13] NMX-AA-079-SCFI-2001. Análisis de Aguas - Determinación de Nitratos en Aguas Naturales, Potables, Residuales y Residuales Tratadas - Método de Prueba. Dirección General de Normas. México, D.F; 2000

[14] UBA. Manual on Methodologies and Criteria for Mapping Critical Levels/Loads and Geographical Areas Where they are Exceeded. UBA Texte 71/96 [Internet]. 1996. http://www.umweltbundesamt.de/mapping/ [Accessed: April 23, 2018]

[15] Matheron G. Principles of geostatistics. Economic Geology. 1963;**58**:1246-1266

[16] NOM-020-SSA1-2014. Salud Ambiental. Valor límite permisible para la concentración de ozono en el aire ambiente y criterios para su evaluación. México, D.F; 2014

[17] NOM-025-SSA1-2015. Salud Ambiental. Valores límite permisibles para la concentración de partículas suspendidas PM_{10} y $PM_{2.}5$ en aire ambiente y criterios para su evaluación. México, D.F; 2015

[18] NOM-021-SSA1-1993. Salud Ambiental. Criterio para evaluar la calidad del aire ambiente con respecto al monóxido de carbono. Valor permisible para la concentración de monóxido de carbono en aire ambiente, como medida de protección a la salud de la población. México, D.F; 1993

[19] Hiltbrunner E, Schwikowski M, Korner C. Inorganic nitrogen storage in alpine snow pack in the Central Alps (Switzerland). Atmospheric Environment. 2005;**39**:2249-2259. DOI: 10.1016/atmosenv.2004.12.037

[20] Fenn M, Geiser L. Temperate Sierra. In: Pardo L, Robin-Abbot M, Driscoll C, editors. Assessment of nitrogen deposition effects and empirical critical loads of nitrogen for ecoregions of the United States. 1st ed. Delaware,OH: United States Department of Agriculture. General Technical Report NRS-80; 2011. pp. 175-180

[21] Grennfelt P, Nilsson J. Critical loads for sulphur and nitrogen. In: Report from a Workshop of Nordic Council Ministers; 19-24 March 1988; Skokloster, Sweden; 1988. 418 p

[22] Escoffie R, Cerón R, Cerón J, Guevara E, Carballo C, Benítez J, et al. Throughfall deposition of N and S to mangrove ecosystems in the Southeast of Mexico. In: Proceedings of the 7th International Conference on Environmental and Geological Science and Engineering (EG'14); 3-5 June 2014; Salerno, Italy. Salerno: WSEAS; 2014. pp. 59-65

[23] Sánchez V. Evaluación del depósito atmosférico de nitrógeno y azufre en la región de Córdoba-Orizaba [thesis]. Ciudad del Carmen Campeche: Universidad Autónoma del Carmen; 2016

[24] López D. Estimación de los flujos de depositación atmosférica de nitrógeno y azufre en la zona metropolitana de la ciudad de Mérida durante 3 temporadas climatológicas [thesis]. Ciudad del Carmen Campeche: Universidad Autónoma del Carmen; 2018

[25] García A, Cerón R, Cerón J, Aguilar C, Montalvo C, Muriel M, et al. Mapping temporal and spatial variation of sulphur and nitrogen deposition to a complex ecosystem in Campeche, México. Transactions on Ecology and the Environment. 2016;**203**:113-123. DOI: 10.2495/EID160111

Chapter 6

Urban Air Pollution Mapping and Traffic Intensity: Active Transport Application

Rasa Zalakeviciute, Adrian Buenaño,
David Sannino and Yves Rybarczyk

Additional information is available at the end of the chapter

http://dx.doi.org/10.5772/intechopen.79570

Abstract

Air pollution represents one of the greatest risks to human health, with most of the world's cities exceeding World Health Organization's recommendations for air quality. In developing countries, a major share of air pollution comes from traffic, consequently, creating air pollution hot spots inside urban street networks. While the world needs to switch to more active and sustainable ways of commuting in order to reduce traffic emissions and help improve degrading cardiopulmonary health due to increasingly sedentary habits, studies point to the negative effects of physical activity near traffic emissions. Common approaches of urban cycling infrastructure planning rely on space availability and route needs, omitting the most vital aspect—air quality. This study, therefore, combines the worldwide need for active commute and health benefits of the cyclists. Our goal was to produce urban pollution map through the geoprocessing of Google Traffic data, validated through the correlation of street level $PM_{2.5}$ (particulate matter <2.5 μm) concentrations and traffic intensity in a selected district of Quito, Ecuador. The multidisciplinary approach presented in this study can be used by city planners all over the world to help identify the cycling network based on air quality conditions and, consequently, promoting active travel.

Keywords: air pollution, urban planning, active travel, mapping, health

1. Introduction

Rapid urbanization, motorization, and industrialization are causing millions of annual deaths by deteriorating air quality—the largest single environmental health risk [1]. In developing world, traffic is one of the major sources of health concerning $PM_{2.5}$ (particles with an aerodynamic diameter less than 2.5 μm) [2]. This demonstrates the widespread need to moderate the rate of motorization by introducing more sustainable ways of transportation. In addition, motorization and technological advances result in diminished physical activity among the world's population, with 23% of adults and 80% of adolescents being insufficiently physically active [3]. This also enhances the risks of cardiovascular diseases, cancer, and diabetes, summing up to one of the leading risk factors for death worldwide [4, 5]. Sadly, obesity levels caused by food and physical inactivity are expected to grow worldwide, and developing countries are projected to exceed the obesity levels of developed countries in the near future [6]. In accordance with the recommendations of World Health Organization (WHO), numerous countries are setting goals on reducing air pollution and insufficient physical activity [3]. It is suspected that both issues of deteriorating air quality and physical activity could be solved by promoting active modes of transportation. In comparison with motorized traffic, cycling alternative requires very little space, and in addition, it is economic, clean, and promotes health [4, 7, 8].

Transport infrastructure is influenced by historical, political, cultural, structural, and economic aspects of urban development [9]. In the era of motorization, the choice of active way of travel over motorized mode depends on a vast range of factors such as proximity, connectivity, convenience, safety, population density, costs, environmental quality, existence of infrastructure and land use mix [10]. Planning urban cycling network is, therefore, a complex task as the existing and potential uses have to be anticipated in order to assure the efficiency of the infrastructure. Simplifying, urban bicycle path planning approaches are based on space availability and route demand by using GIS, GPS, remote sensing and even artificial intelligence techniques [11, 12]. Any new infrastructure, previously unaccounted for in the city's traffic model, has to be incorporated in an existing urban design [8]. In developed countries, efficient regulations are prioritized (traffic calmed residential neighborhoods, car-free city centers, special bicycle streets, etc.) in order to assure the safety of cyclists [9, 13]. At the same time, the route demand requires analyzing the necessity for the bicycle paths. This is done by computational analysis of urban infrastructure and population (census) or user surveys, and recently by mobile phone user applications registering the cycling activity to study the prioritized pathways (e.g., Strava Metro, Kappo, Chaquiñan Urbano, etc.). These applications analyze mobility patterns through voluntary geo-information (crowdsourcing) [14, 15] and interact with the user through games, implemented levels, missions, and rewards to attract new users (Kappo, Capos SpA).

Common approaches, however, often locate the bicycle lanes, tracks, and paths on the roads or in close proximity to motorized traffic. To improve travel safety and accident prevention on the road, a pioneer strategy applies a clear priority rules limited to two types of transport (e.g., bicycle and bus, or bicycle and cars, etc.) [8]. While it helps to reduce the cases of transit accidents, it does little in terms of cardiopulmonary health, especially in developing countries. While earlier studies on cycling focused on engineering, safety, and promoting cycling [7, 16], a number of recent studies discuss health benefits of cycling. For example, research indicates that if using the roads, the personal exposure of cyclists to combustion gas and particle pollution is significantly increased due to the higher concentrations adjacent to traffic and respiratory intensity of cycling [4, 17–22], which is even further increased in high altitude urban areas [23]. In addition, there are no cardiopulmonary health benefits to physical activity on traffic-polluted roads [24]. This study, therefore, proposes a health-centered novel approach to bicycle infrastructure planning based on air quality and traffic intensity. While most of the related research is reported from significantly less traffic-polluted developed countries, this proposal is supported by a study performed in a rapidly growing city of the developing world. Quito is a high elevation (2850 m.a.s.l.), midsize (pop. 2.2 million) city with a decade long $PM_{2.5}$ pollution problem, not only violating WHO recommendations for air quality but even significantly higher limits of national standards [25–27]. Due to older engine technologies (Euro 0–3), high sulfur content fuels (300–650 ppm), terrain inclination and driving style, engine combustion emissions are contributing to over 62% of total $PM_{2.5}$ [28].

This chapter consists of studying the fine particulate pollution at the street level in the central district of Quito, correlating $PM_{2.5}$ concentrations with traffic counts, and comparing the $PM_{2.5}$ pollution map with typical traffic activity, in order to suggest an economic way to plan a healthier active transport infrastructure. Section 2 describes the used methodology. Section 3 is a presentation and a discussion of the results. And Section 4 draws conclusions on the current issues in urban cycling path planning and proposals to improve it.

2. Methods

The historical Mariscal (area 1.5 km^2) district of Quito, Ecuador, was chosen for this study due to the existing network of bicycle lanes and a low variation in traffic intensity. A motorized traffic count was performed to determine the influence of traffic load on the concentration of $PM_{2.5}$. It was evaluated by manual observations [29] based on a week's (February 20–26, 2017) traffic counts, performed in a number of one- and two-way streets of the district. In case of two-way streets, the vehicle count was performed both ways. All heavy (trucks, buses, minibuses) and light (light vehicles and taxis) vehicles were counted from 7 to 8 pm.

At the same time of the experiment, $PM_{2.5}$ pollution scans were performed. To account for different vertical pollution mixing conditions, a multiple coverage of the main streets was performed. Street level $PM_{2.5}$ concentrations were measured using a portable real-time CEL-712 Microdust Pro™ monitor [30] coupled with a GPS [31]. The Microdust Pro was calibrated before the experiment using zero-air and a known concentration filter (164 mg/m^3). The performance of the Microdust Pro sensor based on near-forward angle light scattering technique was validated collocating it with Thermo Scientific 5014i Beta Continuous Ambient Particulate Monitor (R^2 = 0.74). The particle sensor and the GPS were both functioning at a synchronized step of 10 s (particle sensor at 10 s average), held at the height of 1.5 m facing the particle inlet forward while walking at an approximate speed of 2 km/h on the side of the street or an existing bicycle path. The gathered data were then combined with the GPS data to elaborate a pollution map in Qgis. All the collected points were used for the data processing, but atypical data were eliminated. This was done to reduce the impact of these values, which may be the result of some equipment failure [32]. The running average of 1 min was used to represent the points on the map. Geostatistical interpolation of $PM_{2.5}$ concentrations for the district was performed in Qgis using ordinary kriging method. Ordinary kriging is used in pollution dispersion models to estimate an unmeasured region, assuming a constant linear mean over space [33].

Particulate matter $PM_{2.5}$ is mainly produced by the vehicular traffic, especially by diesel vehicles. Thus, in order to assess the correlation between traffic and contamination, pollution maps are compared with traffic maps provided by Google Maps Traffic. Google Maps sets four levels going from fast to slow represented by colors green, orange, red, and brown, respectively. For our analysis, slow traffic representing red and brown were combined together. In the case of $PM_{2.5}$ concentrations, the maximum permissible limits of 24-h international and Ecuadorian regulations were used, three ranges were established: <25, 25–50, and > 50 μg/m^3). These ranges of pollution were also assigned by the same colors: green, orange, and, red respectively; in this case, brown is also combined with red. Once the ranges were defined in the GIS software, in this case Qgis, we proceeded to establish the format of the points for the pixel analysis. It was defined that each spatial point has a size of 6 pixels. In addition, the primary colors of green, blue (accounting for yellow), and red were used for the different levels, because it improved the subsequent computation of the spatial representation of the pollution. To create the traffic layers, the $PM_{2.5}$ pollution map was taken as a basis, and depending on the usual traffic, the color was modified point by point to add the traffic information to each cell. This process was done with the usual traffic of labor days at hours 09h00, 11h00, and 13h00 to cover the hours of the sampling for every studies. A complete street sampling took over 6 h (8h00–14h00) per day.

The original maps produced with Qgis are processed with a program written in Java (**Figure 1**). The first step consists of formatting the maps as a grid. Thus, it is possible to compare the color (level of pollution/traffic) between a cell of the traffic grid and its corresponding cell in the pollution grid. In the second step, the coordinates and the value (1 for green, 2 for yellow, and 3 for red) of each colored cell are recorded in a table, in which the correlation analysis is performed. Two methods are used to carry out the correlation. The first method is based on a strict matching between the colors of the peer cells. Thus, the assessment is Boolean. If the color of both cells is the same, the matching is evaluated as 1, otherwise the value is 0. Then, the value of the correlation (r) is calculated by Eq. (1).

$$r = \frac{\sum_{i=1}^{n}(m_i)}{N} \quad (1)$$

where m_i stands for a true matching and N is the total number of pairs of cells.

The second method is based on a weighted correlation. A weight is calculated according to the amplitude of the difference between the peer cells. For instance, if one cell is red (high) and the other is green (low), the value will be 2. But, if one cell is red and the other is yellow (medium), the value will be 1, only. And if there is no difference between the cells, the score will be 0. For this method, the calculation of the correlation (r) is provided by Eq. (2).

http://dx.doi.org/10.5772/intechopen.79570

$$r = \frac{2N - \sum_{i=1}^{n} | t_i - p_i |}{2N} \tag{2}$$

where t_i stands for the color value of the traffic, p_i stands for the color value of the pollution, and N is the total number of pairs of cells.

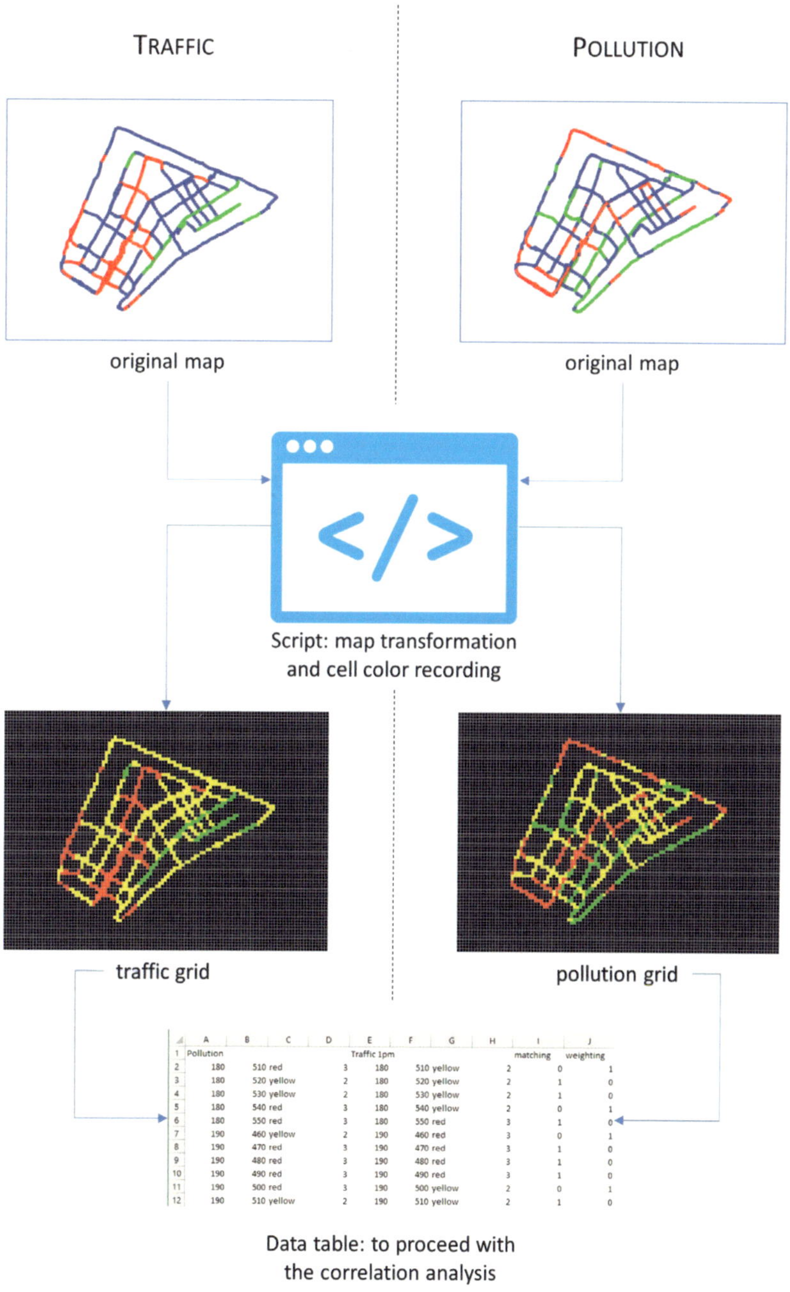

Figure 1. Flowchart to process traffic and pollution data from maps, and compare the values between each matching cell.

3. Results and discussion

During the study, temperature and relative humidity averaged 20°C and 53%, respectively. It is known that higher wind speed helps ventilate the pollution from the street canyons and lower personal exposure [34]. Thus, the low winds registered during the experiment (< 2.5 m/s) offered optimal conditions to measure trapped urban pollution. Higher winds are more common during dry season months of June-August, while current conditions represent Quito weather the rest of the year. This suggests that the results of this study represent usual conditions in the city.

The sampling route (approximately 6 h of sampling and 20 km long) covered a complete urban street infrastructure of the district, which also included sections of four existing bicycle lanes (indicated by broken lines in **Figure 2**). $PM_{2.5}$ concentrations averaged per street and spatial interpolation (ordinary kriging) are represented by the same scale in **Figure 2**. The $PM_{2.5}$ concentrations varied from 27 to 93 μg/m³. These levels exceed the WHO recommended levels for 24-h exposure (25 μg/m³). Meanwhile, punctual $PM_{2.5}$ concentrations (10 s averages) varied from 0 to 624 μg/m³. The sampling peaks nearly exclusively originated from the accelerating diesel buses and minibuses, often at traffic-light-controlled intersections (**Figure 2**).

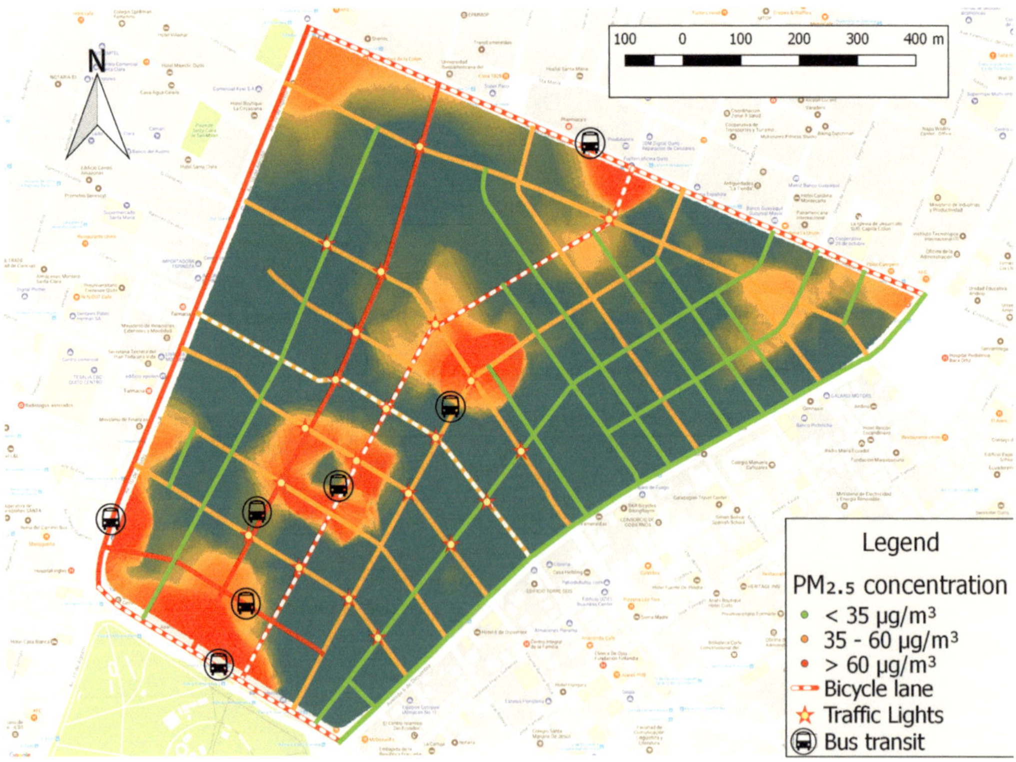

Figure 2. $PM_{2.5}$ concentrations in the urban street network of Quito district Mariscal averaged per street and spatial interpolation (ordinary kriging).

This suggests very high levels of short time exposure to traffic-related $PM_{2.5}$ pollution, which is of a great concern to people near traffic sources.

It can be noted that the concentrations of $PM_{2.5}$ are higher at the streets containing bicycle lanes (**Figure 2**). Statistical analysis of street level concentrations indicated that the average concentration at the streets with cycling lanes is 1.58 times higher than the rest of the streets in the district. This suggests that the personal cyclist exposure to traffic pollution is significantly higher if using existing bicycle lane infrastructure than less densely transited parallel streets, especially streets not permitting city buses (see **Figure 2**). Often, current bicycle paths are located in the most direct and widest streets, thus commonly used by city transport (diesel engines). This places cyclists in the worst air quality conditions suggesting that the users are exposed to the highest level of contamination, and may not be reaping the expected health benefits of active commuting [4, 17–22]. In a few similar studies, the route choices were evaluated in terms of personal exposure to traffic pollution [35, 36], indicating that the choices of "greener" routes significantly reduced personal exposure to direct combustion emissions, but not city background pollution. However, in the case of many cities of the developing world, there are no "greener" options, and the traffic emission levels are significantly higher [37]. Previous studies conclude the importance of the selected travel route, ventilation rate, travel speed for the personal exposure of the person to the traffic pollution, etc. [4, 17]. We confirm the importance of relocating urban bike lanes to the calmer streets, especially in the cities with poor-quality fuels and technologies [17, 20].

For the traffic and $PM_{2.5}$ pollution correlation analysis, the vehicle counts and $PM_{2.5}$ concentration data were averaged per street during all the study period. Correlation analysis is summarized in **Figure 3**. It confirms that there is a strong positive correlation ($R^2 = 0.73$) between the presence of heavy vehicles and the concentrations of $PM_{2.5}$. The number of light personal vehicles and taxis also positively correlated ($R^2 = 0.67$) with the $PM_{2.5}$ concentrations (**Figure 3**). These findings are consistent with other studies [36].

$PM_{2.5}$ concentrations at the street level were also compared with the nearest air quality monitoring station (1.5 km away) representing air quality conditions for central Quito. During the study, the street level pollution was 2.5 times higher than at the monitoring site (elevated at about 10 m above the street infrastructure). The average $PM_{2.5}$ concentrations at the monitoring site were 23.3 ± 8 μg/m^3, while at the street level, the concentrations highly varied at 58.5 ± 91 μg/m^3.There was a positive correlation ($R^2 = 0.42$) between the two measurements, suggesting some relationship between the two traffic-busy areas. However, the significant difference questions current estimates of population exposure to air pollution based on monitoring network data. This especially underestimates the exposure of people that spend a considerable time outside in the street canyons (couriers, police, street vendors, etc.). This inconsistency was suggested by the previous study, where low correlation ($r = 0.31$ all day, $r = 0.49$ morning rush hours) was found between the $PM_{2.5}$ pollution at a monitoring station (elevated above street level) and the surrounding traffic activity [38].

Therefore, a deeper traffic intensity and $PM_{2.5}$ pollution study were performed. We compared the typical traffic at urban street infrastructure with street level $PM_{2.5}$ levels. The results of the

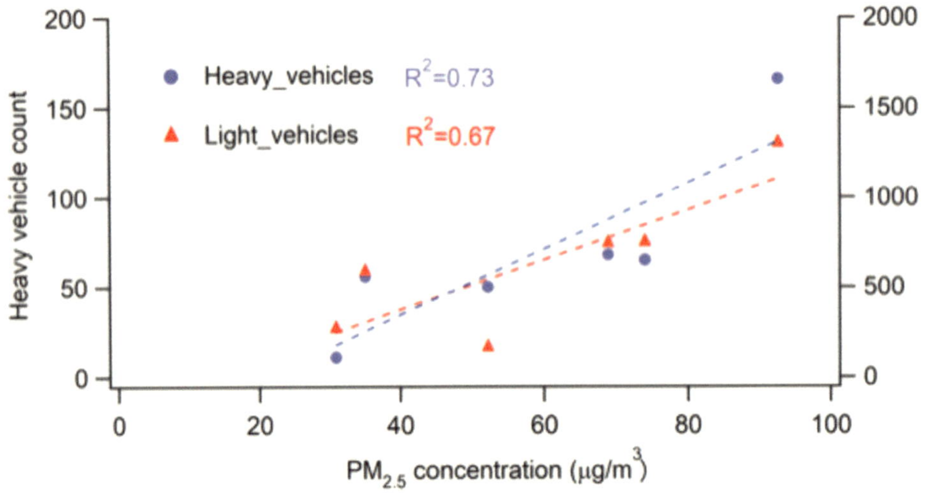

Figure 3. $PM_{2.5}$ concentrations plotted versus the counts of heavy and light vehicles in a few selected main streets, 4/6 of these streets contain bicycle paths (the highest concentrations).

Time	Strict correlation	Weighted correlation
9 am	0.5	0.73
1 pm	0.43	0.69

Table 1. Values of the coefficients r according to the method and the time.

comparison are presented in **Table 1**. The coefficients r obtained from the strict correlation analysis (method 1) are 0.5 and 0.43 at 9 am and 1 pm, respectively. Since the baseline is 0.33 (three possible levels), we can conclude that the correlation is largely above the chance level and, consequently, a significant part of the air pollution is directly explained by the urban traffic. The slight decreases of the correlation in the afternoon can be explained by an augmentation of the dilution of the pollutants in the atmosphere that occurs at this time of day [38]. These results are confirmed by the weighted correlation analysis. This second method provides us with coefficients of 0.73 and 0.69 at 9 am and 1 pm, respectively. Although the baseline of this method is higher (r = 0.5) than in the first analysis, the obtained values cannot be the effect of the hazard. As expected, the correlation between traffic and $PM_{2.5}$ is superior when the concentrations are recorded at the street level than at the monitoring station level. Taken together with the results presented in **Figure 3**, these findings support the hypothesis of considering traffic density in the planning of urban cycling paths.

While there is a serious traffic and physical inactivity problem in the world, one of the seemingly best solutions—cycling—might not be widely adopted due to multiple issues such as missing infrastructure, crime/safety, and environmental pollution [39]. The results of this study encourage city planners to locate cycling paths on less trafficked, light vehicle streets

http://dx.doi.org/10.5772/intechopen.79570

rather than on major streets, especially in developing countries using high sulfur content fuels that cause more particulate pollution. Not many cities can afford greenways (undeveloped land in or near urban area) for cycling; thus for the best solution, some lighter traffic density parallel street options can be used to redirect bicycle traffic to reduce the exposure to high concentrations of primary pollutants. Following the example of Amsterdam, Netherlands, the bicycle paths could be located on the streets of exclusively light vehicle traffic, not only reducing the risks of safety but also air pollution. This could further encourage new conversions toward more active commuting.

4. Conclusions

To the best of our knowledge, this is the only study proposing to base urban cycling path planning on the benefits of cardiopulmonary health and offering an economic solution applicable for any country. During the study, the street level pollution in a central district of Quito was 58.5 ± 91 $\mu g/m^3$, significantly exceeding the WHO recommended levels for air quality. This large variation suggests an extremely high level of short time exposure to traffic-related $PM_{2.5}$ pollution, which is of a great concern to people near traffic sources. The results of this study show that there is a strong positive correlation between the amount of heavy diesel vehicles (especially city busses) on the road and the concentrations of $PM_{2.5}$. We also demonstrate that most of the bicycle paths in the central Quito are located on the most polluted streets. This indicates the importance of an appropriate selection of routes with low vehicular traffic load to reduce cyclists' exposure to fine particulate matter. We also conclude a high correlation between the motorized traffic intensity (Google Traffic Maps service) and $PM_{2.5}$ pollution. Traffic maps offer a reliable and economic method for healthier cycling infrastructure planning in any city of the world. Therefore, this study serves as a reference for implementing control measures for public transport and for the planning of strategic routes, as well as the implementation of adequate infrastructure to support active transportation by reducing vehicular pollution exposure and promoting human health.

Acknowledgements

We want to thank the Secretariat of the Environment of the Municipality of the Metropolitan District of Quito in Quito, Ecuador, especially Maria Valeria Diaz Suarez for the infinite collaboration.

Conflict of interest

The authors declare no conflicts of interest.

Author details

Rasa Zalakeviciute[1*], Adrian Buenaño[1,2], David Sannino[3] and Yves Rybarczyk[1,4]

*Address all correspondence to: rasa.zalake@gmail.com

1 Universidad de Las Americas, FICA, Intelligent & Interactive Systems Lab (SI^2 Lab), Quito, Ecuador

2 Department of Environmental Quality, Ministry of Environment, Porto Viejo, Ecuador

3 Department of Microbiology, Cornell University, Ithaca, NY, USA

4 Universidade Nova de Lisboa, DEE/UNINOVA/CTS, Monte de Caparica, Portugal

References

[1] WHO. WHO I 7 Million Premature Deaths Annually Linked to Air Pollution Page 1 of 2 WHO I 7 Million Premature Deaths Annually Linked to Air Pollution Page 2 of 2. Media Cent; 2014. pp. 6-7

[2] Karagulian F et al. Contributions to cities' ambient particulate matter (PM): A systematic review of local source contributions at global level. Atmospheric Environment. 2015;**120**:475-483

[3] WHO. WHO I Physical Activity. WHO; 2018. Retrieved from: http://www.who.int/en/news-room/fact-sheets/detail/physical-activity

[4] de Hartog JJ, Boogaard H, Nijland H, Hoek G. Do the health benefits of cycling outweigh the risks? Environmental Health Perspectives. 2010;**118**(8):1109-1116

[5] Bauman AE. Updating the evidence that physical activity is good for health: An epidemiological review 2000-2003. Journal of Science and Medicine in Sport. 2004;**7**(1):6-19

[6] Bhurosy T, Jeewon R. Overweight and obesity epidemic in developing countries: A problem with diet, physical activity, or socioeconomic status? Scientific World Journal. 2014. 7 p. Article ID: 964236. http://dx.doi.org/10.1155/2014/964236

[7] Handy S, van Wee B, Kroesen M. Promoting cycling for transport: Research needs and challenges. Transport Reviews. 2014;**34**(1):4-24

[8] van der Horst I. Cycling Culture in Amsterdam. 2014

[9] Koglin T. Velomobility and the Politics of Transport Planning. 2014

[10] Saelens BE, Frank LD, Med AB. Environmental correlates of walking and cycling: Findings from the transportation, urban design, and planning literatures. Society of Behavioral Medicine, no. c. 2003:80-91

[11] Guerreiro T d CM, Kirner Providelo J, Pitombo CS, Antonio Rodrigues Ramos R, Rodrigues da Silva AN. Data-mining, GIS and multicriteria analysis in a comprehensive

method for bicycle network planning and design. International Journal of Sustainable Transportation. 2018;**12**(3):179-191

[12] Forsyth A, Krizek K. Urban design: Is there a distinctive view from the bicycle? Journal of Urban Design. 2011;**16**(4):531-549

[13] Pucher J, Buehler R. Making cycling irresistible: Lessons from the Netherlands, Denmark and Germany. Transport Reviews. 2008;**28**(4):495-528

[14] Orellana D, Hermida C, Osorio Guerrero P. Comprendiendo los patrones de movilidad de ciclistas y peatones. Una síntesis de literatura. Revista Transporte y Territorio. 2017;**16**:167-183

[15] Strava. Data-Driven Bicycle and Pedestrian Planning. Published at Hubspot; 2016. pp. 1-4. Retrieved from: http://cdn2.hubspot.net/hubfs/1979139/Strava_Metro___Data-Driven_Planning.pdf

[16] Pucher J, Dill J, Handy S. Infrastructure, programs, and policies to increase bicycling: An international review. Preventive Medicine (Baltimore). 2010;**50**:S106-S125

[17] Apparicio P, Carrier M, Gelb J, Séguin AM, Kingham S. Cyclists' exposure to air pollution and road traffic noise in central city neighbourhoods of Montreal. Journal of Transport Geography. 2016;**57**:63-69

[18] Hankey S, Marshall JD. On-bicycle exposure to particulate air pollution: Particle number, black carbon, PM2.5, and particle size. Atmospheric Environment. 2015;**122**:65-73

[19] Int Panis L et al. Exposure to particulate matter in traffic: A comparison of cyclists and car passengers. Atmospheric Environment. 2010;**44**(19):2263-2270

[20] MacNaughton P, Melly S, Vallarino J, Adamkiewicz G, Spengler JD. Impact of bicycle route type on exposure to traffic-related air pollution. The Science of the Total Environment. 2014;**490**:37-43

[21] Quiros DC, Lee ES, Wang R, Zhu Y. Ultrafine particle exposures while walking, cycling, and driving along an urban residential roadway. Atmospheric Environment. 2013;**73**:185-194

[22] Wu DL et al. Influences of commuting mode, air conditioning mode and meteorological parameters on fine particle (PM2.5) exposure levels in traffic microenvironments. Aerosol and Air Quality Research. 2013;**13**(2):709-720

[23] Bravo Alvarez H, Sosa Echeverria R, Sanchez Alvarez P, Krupa S. Air quality standards for particulate matter (PM) at high altitude cities. Environmental Pollution. 2013;**173**:255-256

[24] Sinharay R et al. Respiratory and cardiovascular responses to walking down a traffic-polluted road compared with walking in a traffic-free area in participants aged 60 years and older with chronic lung or heart disease and age-matched healthy controls: A randomised, crosso. Lancet. 2018;**391**(10118):339-349

[25] Kleine Deters J, Zalakeviciute R, Gonzalez M, Rybarczyk Y. Modeling PM2.5 urban pollution using machine learning and selected meteorological parameters. Journal of Electrical

and Computer Engineering. 2017. 14 p. Article ID: 5106045. https://doi.org/10.1155/2017/5106045

[26] Zalakeviciute R, Rybarczyk Y, López-Villada J, Diaz Suarez MV. Quantifying decade-long effects of fuel and traffic regulations on urban ambient PM2.5 pollution in a mid-size south American city. Atmospheric Pollution Research. 2017;**9**(1):66-75

[27] Rybarczyk Y, Zalakeviciute R. Machine learning approach to forecasting urban pollution. In: 2016 IEEE Ecuador Technical Chapters Meeting; ETCM; 2016

[28] Secretaría de Ambiente. Secretaria de Ambiente: Informe Final Inventario de Emisiones de Contaminantes Criterio, DMQ 2011. 2014. p. 53

[29] Skszek S. Report on Non-Traditional Traffic Counting Methods. Vol. 85748. Tanque Verde Loop Rd. Tucson, AZ: 2001. p. 505

[30] Casella. Microdust Pro Real-Time Dust Monitor. No. 800. pp. 1-62

[31] Garmin. eTrex Owner's Manual. 2011. p. 60

[32] Rohde RA, Muller RA. Air pollution in China: Mapping of concentrations and sources. PLoS One. 2015;**10**(8):e0135749

[33] Kim S-Y et al. Ordinary kriging approach to predicting long-term particulate matter concentrations in seven major Korean cities. Environmental Health and Toxicology. 2014;**29**:e2014012

[34] Kingham S, Meaton J, Sheard A, Lawrenson O. Assessment of exposure to traffic-related fumes during the journey to work. Transportation Research Part D: Transport and Environment. 1998;**3**(4):271-274

[35] Hertel O, Hvidberg M, Ketzel M, Storm L, Stausgaard L. A proper choice of route significantly reduces air pollution exposure—A study on bicycle and bus trips in urban streets. The Science of the Total Environment. 2008;**389**(1):58-70

[36] Jarjour S et al. Cyclist route choice, traffic-related air pollution, and lung function: A scripted exposure study. Environmental Health. 2013;**12**(1):14

[37] WHO | Air pollution levels rising in many of the world's poorest cities. WHO: 2017

[38] Rybarczyk Y, Zalakeviciute R. Regression models to predict air pollution from affordable data collections. In: Machine Learning—Advanced Techniques and Emerging Applications. 2018

[39] Majumdar BB, Mitra S. Development of level of service criteria for pedestrians. Journal of Urban Planning and Development. 2018;**144**(2)

Chapter 7

Dioxins and Furans: Emerging Contaminants of Air

Muhammad Zubair and Amina Adrees

Additional information is available at the end of the chapter

http://dx.doi.org/10.5772/intechopen.80680

Abstract

Dioxins and furans are classified highly contaminating toxic chemicals having serious effect on human health. This chapter begins with a brief summary on the formation, occurrence and toxicity of polychlorinated dibenzo-p-dioxins (PCDDs) and polychlorinated dibenzofurans (PCDFs) with their potential health effects, mitigation measures of these harmful compounds. Depending on position of chlorine atoms on aromatic rings, about 210 chemically different PCDD/PCDFs known as "congener" are present in the environment. The expected biological activity of PCDD/PCDFs is expressed relatively to the activity of 2,3,7,8-tetrachloro-dibenzo-p-dioxin (TCDD) using a common metric. TCDD is the most biologically potent among all the congeners. Toxic level of a mixture of PCDD/PCDFs is therefore expressed in TCDD toxicity equivalents or TEQs. There are two mechanisms for the formation of dioxins and furans, one from precursors and other by de novo synthesis. PCDD/PCDFs followed a mechanism which uses macromolecule carbon and chlorine in fly ash to form dioxins at low temperature. There are various sources of the formation of PCDD/PCDFs like hospital waste incinerators, industrial combustion and burning of domestic waste. Dioxins and furans have very harmful effects on the human health causing cancer, diabetes, neurotoxicity, immunotoxicity and chloracne. It has been experienced that hybrid method secure a sustainable future for the incinerators and PCDD/PCDFs removal technologies. This chapter will help the researchers and practitioners for better understanding and decision making for future research to establish a sustainable PCDD/PCDFs free life.

Keywords: dioxin, furan, contaminant, toxicity, inciration, mitigation

1. Introduction

Emerging contaminants are those which were not considered as such as in previous times, but exist in the environment on global level. They are common derivatives of municipal, agricultural and industrial sources and pathways. A considerable rise has been observed in terms of emission of air pollutants in the atmosphere [1, 2]. Environmental quality may be at risk by developmental activities as imbalance in the composition of air. Persistent organic pollutants (POPs) are regarded as very harmful compounds because they are resistant to various factors of biochemical and photolytic degradation. POPs are persistent to soils, sediments, and air for several decades [3]. Having high toxicity and long persistency in the environment they accumulate in the fatty tissues of humans and animals resulting into many behavioral, reproductive and developmental changes [4, 5]. POPs have gained global attention due to their transportation over long distances from the source. Various researchers proved that persistent organic pollutants (POPs) are dangerous compounds due to their persistent, bio accumulative and toxic characteristics. POPs include pesticides, polychlorinated biphenyls, polycyclic aromatic hydrocarbons (PAHs), polychlorinated dibenzo-p-dioxins (PCDDs), polychlorinated dibenzofurans (PCDFs), and groups of brominated flame retardants [6]. In Stockholm convention on POPs it was decided to control the concentration of the dirty dozen in the environment. PCDDs and PCDFs are members of dirty dozen which are classified as toxic and carcinogenic, found in very small amounts in the environment. The major issue of PCDDs/PCDFs is due to their extreme persistency in the environment causing toxicity and cancer to living organisms and can potentially cause cancer. Dioxin and furans are family of chlorinated hydrocarbon compound which are categorized into three main classes as: polychlorinated dibenzo-p-dioxins (PCDDs), polychlorinated dibenzofurans (PCDFs) and dioxin-like polychlorinated biphenyl (DL-PCBs). PCDDs and PCDFs are produced from different anthropogenic activities like forest fires, domestic and hospital waste incineration [7]. They are by products of the synthesis or combustion of chlorine based compounds that include some of the most toxic chemical substrates.

Polychlorinated dibenzo-p-dioxins (PCDD), polychlorinated dibenzofurans (PCDF) are produced accidentally due to inadequate combustion, as well as during the manufacturing and formulation of chlorinated pesticides and other substances. They are also emitted from the incineration of hospital, municipal and hazardous waste. There are seventy many dioxins, out of which seven are considered most toxic to humans, aquatic and terrestrial organisms, causing congenital mental retardation (endocrine disrupting) and physical disorders. Many industrial processes are likely to generate a huge amount of industrial waste, which is openly burned without any safety measure, producing huge quantity of Polychlorinated dibenzo-p-dioxins (PCDD) and polychlorinated dibenzofurans (PCDF).

The polychlorinated dibenzo-p-dioxins (PCDDs) and dibenzofurans (PCDFs) are two groups of tricyclic, planar, aromatic, nonpolar, poorly water soluble, lipophilic and stable chemicals [8]. Out of 210 congener of these chemicals 17 are more toxic. These are toxic carcinogenic unintentional by-products which are found in lesser concentration in environment, reflecting drastic effect due to their extreme persistency. Their persistency is due to their long half-life of 10–20 years roughly [7].

1.1. Environmental concentration

Emission favors the formation, environmental release and distribution of these congeners of TCDDs and TCDFs signatures of the types of dioxin and furans associated with particular

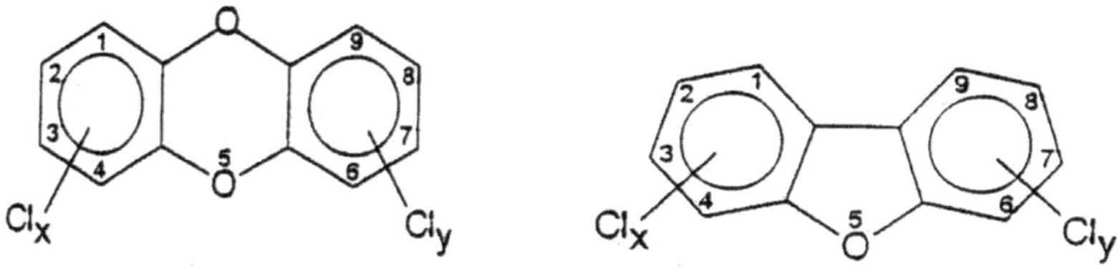

Figure 1. General structures of dioxin and PCDDs (left) and furan and PCDFs (right).

environmental sources of these compounds [9]. On the contrary, the PCDD/Fs in the ambient air samples, characterized by the abundance 76% in some parts of the world and it is dominated by 1,2,3,4,6,7,8-HpCDF (40%) followed by OCDD (10%), OCDF (6%), and 1,2,3,4,6,7,8-HpCDD (11%). Reported values not similar to any of the profiles reported by Cleverly; however, it reveals the simultaneous occurrence of two potential emission sources of TCDDs and TCDFs [9].

1.2. Dioxins

Dioxins chemically consist of two benzene rings, connected by a pair of oxygen atoms. Each of the eight carbon atoms on the rings that are not bonded to oxygen can bind with hydrogen atoms or atoms of other elements. The more toxic dioxin is one having chlorine atoms at the 2, 3, 7, and 8 positions. This isomer 2,3,7,8-TCDD is chemically most stable and toxic compound. Being Water insoluble dioxin is not diluted with rain water and remain in the soil thus penetrate and accumulate in fatty tissues. The molecular structure of dioxins can be determined by single crystal X-ray diffraction crystallography. The structure of TCDD in crystal which is also observed in molecular crystal of poly aromatic hydrocarbon (PAHs) [10].

1.3. Furans

Furans are also toxic organic compounds which are colorless, highly flammable and very low boiling point almost equal to room temperature. Furans are heterocyclic compound containing one oxygen atom, four carbon atoms and a 5-membered ring (**Figure 1**). The global environmental impact of volatile compounds depends on many factors such as air pressure, temperature and also the weather and terrain features which affect the deposition. It is interesting to know what derivatives are formed from reactions with furan, and also how fastly these potential products are removed from the environment. One of the 5 largest sinks of furan derivatives will probably be reaction with the hydroxyl radical (OH), which mostly initiates reactions with VOCs through hydrogen abstraction from a C-H bond.

2. History of dioxins and furans

The environmental distribution of dioxins and furans is a function of transport and source. The sources of dioxins and furans are combustion, soil deposition, volatilized and transported particulates which were sequestered and rereleased into the environment. It is evident that environmental dispersal and accumulation of these compounds is not necessarily depending upon a nearby source. Polychlorinated dibenzo-p-dioxins (PCDDs) and their cousins, the polychlorinated dibenzofurans (PCDFs), are notorious environmental contaminants. Depending on the position of chlorine atoms attached on ring, two hundred and ten chemically different toxic compounds of dioxins and furans are produced each of which is known as "congener" [11]. Collectively these congers or compounds are often known as "dioxins and furans". These compounds have received considerable public and scientific attention because of their acute toxicity. Out of all these congeners, 2,3,7,8-TCDD has lowest known LD_{50} values. It takes only 0.6 μg/kg of body weight to kill male guinea pigs [12]. In 1957, a strange disease killed millions of young chickens in the eastern and mid-western U.S. and symptoms of this disease were excess of fluid in the heart sac and abdominal cavity, chased to the fatty acids that had been added to the chicken's feed. Efforts of several years lead to the isolation of one of the identified toxic chemical [13] by X-ray crystallography; it was 1,2,3,7,8,9-hexachlorodibenzo-p-dioxin. In the 1960s and early 1970s, the Northeastern Pharmaceutical and Chemical Company (NEPACCO) established a plant in Veron for manufacturing of hexachlorophene from 2,4,5-trichlorophenol and formaldehyde. Unfortunately, 2,3,7,8-TCDD was an impurity in the 2,4,5-trichlorophenol starting material used in this process; thus, the hexachlorophene product needed to be purified before sale. Because of its neurotoxicity, the U.S. Food and Drug Administration restricted the use of hexachlorophene in 1971. In 1977 Olie et al. investigated dioxins which existed in fly ash from an industrial heating facility [14]. In 2000, Bumb et al. in a famous paper "Trace chemistries of fire: A source of chlorinated dioxins", exposed that dioxins were exist in particles from the combustion of organic material, involving the combustion of municipal and chemical waste. This was marvelous discovery, suggested that "dioxins have been with us since the advent of fire" [15, 16]. In mid of 1970, a Swiss company established a chemical plant for manufacturing of 2,4,5-trichlorophenol by the reaction of 1,2,4,5-tetrachlorobenzene with NaOH. Unfortunately, accident occurred and chemicals from vessels were released and transported by wind. This caused a lot of disaster to plants, animals and human. Later on it was confirmed that the reason of this disaster was a notorious chemical 2,3,7,8-TCDD [17]. During war in Vietnam, US military showered Agent Orange from 1965 to 1971 as defoliant to kill food crops. Agent Orange was a mixture of n-butyl esters of 24-D and 245-T, the latter of which was formed from 2,4,5-trichlorophenol. Agent Orange was contaminated with small amounts of 2,3,7,8-TCDD [18]. Dioxins and furans including dioxin-like polychlorinated biphenyls collectively as known as DLCs. These organic compounds are highly toxic and accumulate, through the food chain, into the lipid component of animal foods. In another study it has been observed that levels of these compounds in the environment are declining since 1970s [19]. Highly exposed groups of dioxins and furans are found in breastfeeding infants, fishers and workers of cement industries. Although Dioxins and furans have been extensively studied as a contaminant, but still a great deal of research is needed regarding their potential toxicity.

3. Toxicity of dioxins and furans

The toxicity and exposure preferably depends on the composition and particle size of the mixtures containing toxic compounds. Biological activity of mixtures of dioxins and furans is desirable to express the common effect. Literature review reports the biological activities

CAS number	Hazardous substance	Mammalian TEF	Avian TEF
Dioxins			
1746-01-6	2,3,7,8-Tetrachloro dibenzo-p-dioxin	1	1
40321-76-4	1,2,37,8-Pentachloro dibenzo-p-dioxin	1	1
39227-28-6	1,2,3,4,7,8-Hexachloro dibenzo-p-dioxin	0.1	0.05
57653-85-7	1,2,3,6,7,8-Hexachloro dibenzo-p-dioxin	0.1	0.01
19408-74-3	1,2,3,7,8,9-Hexachloro dibenzo-p-dioxin	0.1	0.1
35822-46-9	1,2,3,4,6,7,8-Heptachloro dibenzo-p-dioxin	0.01	<0.001
3268-87-9	1,2,3,4,6,7,8,9-Octachloro dibenzo-p-dioxin	0.0003	0.0001
Furans			
51207-31-9	2,3,7,8-Tetra chloro dibenzofuran	0.1	1
57117-41-6	1,2,3,7,8-Pentachloro dibenzofuran	0.03	0.1
57117-31-4	2,3,4,7,8-Pentachloro dibenzofuran	0.3	0.1
70648-26-9	1,2,3,4,7,8-Hexachloro dibenzofuran	0.1	0.1
57117-44-9	1,2,3,7,8-Hexachloro dibenzofuran	0.1	0.1
72918-21-9	1,2,3,7,8,9-Hexachloro dibenzofuran	0.1	0.1
60851-34-5	1,2,3,4,6,7,8-Heptachloro dibenzofuran	0.01	0.01

Table 1. TEQ values for mammalians and Avians (EPA, 2000).

of the various dioxin and furans congeners comparing with of 2,3,7,8-tetrachloro-dibenzo-p-dioxin (TCDD). Information has been gathered for this dioxin and furan congener, and it is found that TCDD is the most biologically toxic among the mixtures of dioxins and furans and highly potent. The toxicity of TCDD/TCDFFs is therefore expressed in toxicity equivalents (TEQs) of highly potent component as TCCD (**Table 1**). As an example, exposure to a matter a potency of 2.0 ng TEQ/kg means that matter has the potency equal to 2.0 ng TCDD/kg. Toxicity equivalents values for various materials are calculated by multiplying the mass or concentration of each component by a TEF and adding all present components. The biological activities of TCDD/TCDFs vary depending on the human's exposure. Even so, public concern persists regarding food supply and adverse outcomes to TCDD/Fs exposure, especially in highly exposed populations. Sensitive population includes fetuses and new born infants. These populations may be at increased risk due to exposure through foods. However, many foods which are sources of dioxins and furans are also sources of important nutrients, such as calcium (Ca) and vitamins (A E C), protein, iron and fish.

4. Chemical formation and sources of dioxins and furans

PCDDs and PCDFs are not produced intentionally but as by product of various processes like chlorinated phenols, PCBs, phenoxy herbicides, chlorinated benzene, chlorinated aliphatic compounds, chlorinated catalysts and halogenated biphenyl ethers **Table 2** [20, 21]. There are

Equation no.	Reaction steps	Rate constants
R1	P=P. +H	$10^{15.5}$ exp[−86,500/(RT)]
R2	P + OH = P. + H_2O	10^9 exp [−86,500/(RT)]
R3	P. = Pr	$10^{13.6}$ exp[−57,654/(RT)]
R4	P + P. = PD + Cl	10^9 exp[−26,000/(RT)]
R5	PD = D + HCl	10^{14} exp[−45,000/(RT)]
R6	PD + OH = D + H_2O	10^9
R7	P. + R = P + R.	10^8 exp[−26,000/(RT)]
R8	P. + OH = Pr	10^9
R9	D = Pr	$10^{15.5}$ exp[−80,000/(RT)]
R10	D + OH = Pr	10^8
R11	P. + O_2 = Pr	10^8
R12	R + OH = R. + H_2O	10^9
R13	R = Pr	10^{18} exp[−90,000/(RT)]

P = polychlorinated phenols, P. = polychlorinated phenoxy radicals, Pr = unspecified products, PD = polychlorinated 2-phenoxy phenols, D = PCDD, R = fuel molecules.

Table 2. Mechanism of PCDD/PCDFs formation according to Shaub and Tsang.

two mechanisms for the formation of PCDD/PCDFs from solid waste incinerator: formation from precursors and formation by de novo synthesis. PCDD/PCDFs followed a mechanism which uses macromolecule carbon and chlorine in fly ash to form dioxins at low temperature by involving the oxidative breakdown and conversion of macromolecular carbon structure to aromatic compound [22]. PCDD/PCDFs formed by this process have solid phase at one end and other part will desorb to gas phase and be carried by off-gas flow. Dioxin can be formed by variety of precursors like chlorobenzene and chlorophenol. Such precursors are produced by partial combustion or heterogeneous catalytic reaction on surface of fly ash [23, 24]. The examination of human tissue of earlier time showed little concentration of dioxins than today [25]. Analysis of sediment near industrial area showed that dioxin concentration is minor till 1920 [8]. However its concentration continued to increase from 1920 to 1970 [26].

4.1. Incineration sources

The most effective means of dealing with the problem is to reduce the amount of wastes generated by hospitals. Incineration is one of the key methods to reduce the amount of generated waste. But one serious drawbacks of this process is the emission of dioxins and furans in flue gas. Dioxins and furans are contaminants that are released into the environment from combustion processes. The combustion of plant material from forest, brush, and range fires contributed to preindustrial deposition of dioxins into soil, sediment, and clay. Postindustrial sources are varied and include industrial burning (e.g., steel, coke, ceramic, and foundry), landfill fires, structural fires, utility pole and transformer storage yards, crematories, and backyard barrel burning of trash and woody and other plant material. Municipal solid waste incinerator produced massive amount of dioxins by following the precursors and de novo synthesis mechanism [27]. Past few decades demonstrated that formation of dioxin and dioxin like compound from municipal solid waste incinerator is about 50 ng 1-TEQ/kg [28]. Hospital wastes are also treated by incineration without using high quality technologies. Thus incineration of chlorine containing product produced a lot of dioxins to atmosphere [28, 29]. Hazardous waste (explosive, oxidizing, highly inflammable, infectious, mutagenic) incinerations are responsible for the production of dioxin [30]. Solid residues produced from waste water treatment containing toxic organic pollutants are called sewage sludge. Limitation to Landfill disposal process it is also treated by incineration. Thus it also a big source of dioxins [31].

4.2. Industrial sources

Use of harmful material as fuel leads to major contribution toward dioxin and furans in clinker goes to air [32]. A number of studies of emission confirmed that wood burning is a major source of TCCD/PCDFs emission in air [33]. One of review that give appealing information is that dioxin emission from wood burning is 945 g I-TEQ per year [34]. Different researchers from Norway and Sweden have studied dioxin emission from vehicles engines combustion of fuels [35]. Waste products released from paper and pulp mills have large no of chlorinated and phenolic compounds as in wood pulps leads to emission of dioxin in water, land and paper product [36]. One report from china suggested that paper industry producing 300 pg/11-TEQ [37]. Metallurgical industries involving a no. of process like smelting operation and scrap metal recovery are major sources of dioxin and similar congeners [38]. Sintering of iron ore is also a source of these toxic congeners [39]. Annual production of dioxin in world is estimated to be 500–400 g I-TEQ.

4.3. Reservoir sources

PCDDs and PCDFs are persistent and water insoluble compounds having high tendency to accumulate in soil and sediment; vegetation, waste and in organic matter. These compounds have the ability to be recycled and redistributed in environment. Biological processes are also a source of dioxin and furans. When microorganisms catalyze chlorinated phenolic compounds then there occurs the emission of toxic compounds of dioxins and furans [40]. There are no miscellaneous sources e.g. power generation, thermal oxygen cutting metal at demolition sites, Kraft Liquor boiler, laboratory waste tire combustion, carbon activation services [41].

5. Effects of dioxins and furans

Comprehensive study has been carried out on toxicity of TCDD/PCDFs and its related compound [42]. Rodents when exposed to TCDD/Fs it lessened the reproductive capability of female and disrupt the sperm production in male progeny. Many diseases like hypospadias, ectopic testes, vaginal pouches, agenesis of the ventral prostate, and nipple retention were notice [43]. Exposure to TCDD leads to prevalence and complication endometriosis [44]. Dioxin and furans are well declared endocrine disrupters thus lessened the production of thyroid hormones [45–47]. Exposure of wildlife to dioxins cause many reproductive variations such as cryptorchidism in the Florida panther, small baculum in young male otters, small penises in alligators, sex reversal in fish, and altered social behavior in bird [45].

TCDD/PCDFs are recommended highly damaging to immune system and thus decreasing host resistance to infectious diseases and lowered immune responses. Dioxins also disturb production of inflammatory cytokines such as interleukin and necrosis factor [48]. TCDD/PCDFs significantly affect the neuron populations of vertebrate brain; however their damage to brain function is still not clear and need more research to reveal the truth. It is noticed that TCDD/PCDFs affect the gonadal and thyroid hormones and slow down the neural transmission network [49].

Chloracne is a skin damaging condition with both hyper keratotic and hyper proliferative responses of the epidermis is caused by the exposure of TCDD/Fs. Many animals such as cows, horses and rabbits also revealed this disease notice [50]. In addition to these, loss of sebaceous gland and atrophy of hair follicle also noticed due the severe exposure of TCCD/PCDFs [51]. With the exposure to TCDD/PCDFs excessive keratinization may also occur in dermal epithelium [52].

Some biochemical changes also seen with exposure to congeners of TCDD/Fs [53]. It was noticed that insulin level goes down after TCDD/PCDFs massive accumulation. It is also observed TCCD/PCDFs exposure effect the body growth, deplete the energy stores and thus organism has to lower insulin level to sustain blood glucose levels. On the other side tryptophan concentrations of brain increase, due to increase free fatty acids in blood circulation. These changes allow tryptophan to compete with binding site of albumin and help its transport to central nervous system [54]. Similarly, oxidative processes have been considered necessary for metabolic (e.g., porphyria) and morphological damage of the liver [55].

Another study reports that TCCD/PCDFs have effects on female Rhesus monkeys when exposed to 0, 5, or 25 ng/L in their diet for 4 years and then protected for 10 additional years. It was noticed that these monkeys caught with severity of endometriosis [56]. On the other hand, endometriosis cyst growth in both rats and mice has also been enhanced by exposure to TCDD/PCDFs at very low doses where no ovarian toxicity occurred [57]. However, a dose of 10 mg TCDD/kg for a 16 week period resulted in ovarian atrophy in mice [58] .

6. Mitigation and control of dioxins and furans

Various techniques are available for the controlled emission of dioxins and furans depending upon the type of feed stocks. Some popular and effective methods are: using sulfur compounds namely $(NH_4)_2SO_4$, pyrite (FeS_2), changing the operating conditions of incinerations.

Present chapter discuss methods to minimize effect of dioxins and benzofurans and their formation in different types of incineration systems. Municipal solid waste incineration system, hazardous solid waste incineration system and bio medical waste incineration system. Formation mechanism and the various sources including the precursors of PCDD/Fs formation need to be controlled during the combustion. It is also evident through literature that risk management strategies to reduce polychlorinated dibenzo-dioxins (PCDDs) and dibenzofurans (PCDFs) exposure, consideration may be given to the potential impact of changes to food and nutrition policies, particularly those related to public health and food.

There are number of technologies practiced around the globe to control combustion practice coupled with end of pipe treatment. Most preferably these are selective catalytic reduction, addition of suitable inhibitors for dioxins and furans eradication in the flue gas of waste incinerators. A brief background with pictorial images of the present technologies has been presented in this chapter (**Figure 2**) three different technologies used for the control of dioxins and furans have been discussed. Selective catalytic oxidation or reduction (SCR) using NH3-SCR catalysts (commercial V_2O_5-WO_3/TiO_2) could effectively promote the decomposition of PCDD/PCDFs without hampering the normal operating conditions of the incineration studied by [60]. Vermeulen et al. added urea for the purpose of decomposition of dioxins reducing it upto 90% under the same operating conditions as of ammonia [59]. In the process of adding compounds of sulfur to control the formation different congeners of PCDD/PCDFs (**Figure 3**) the feedstock is firstly prepared, homogenized with crushing and adding into the rotary kiln. The emission of dioxin compounds were controlled by a series of operations that includes quenching tower, acid neutralizing tower, wet scrubber, bag filter and activated

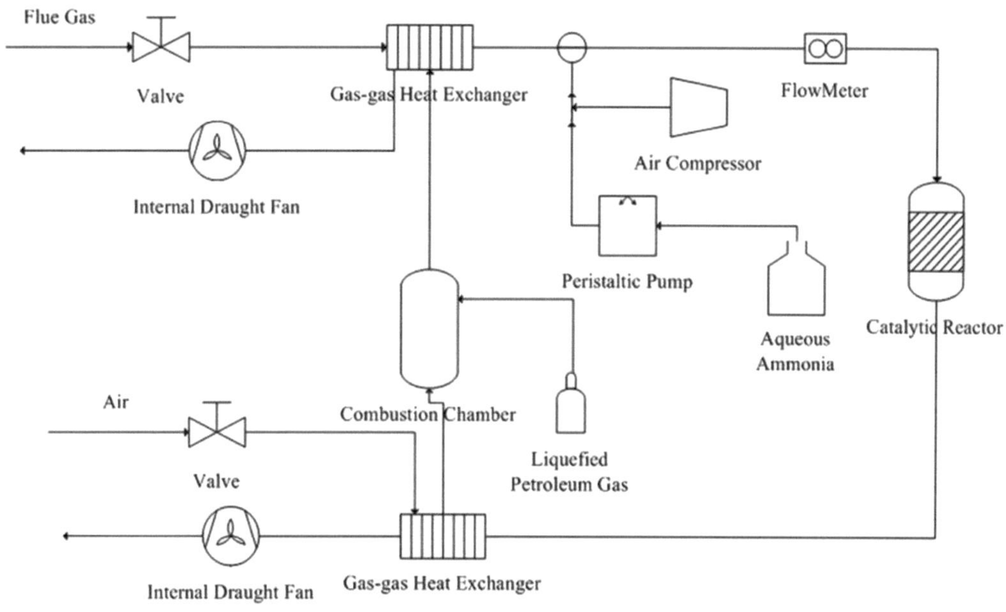

Figure 2. Inhibition of PCDD/PCDFs by the aid of V_2O_5-WO_3/TiO_2 catalysts [59].

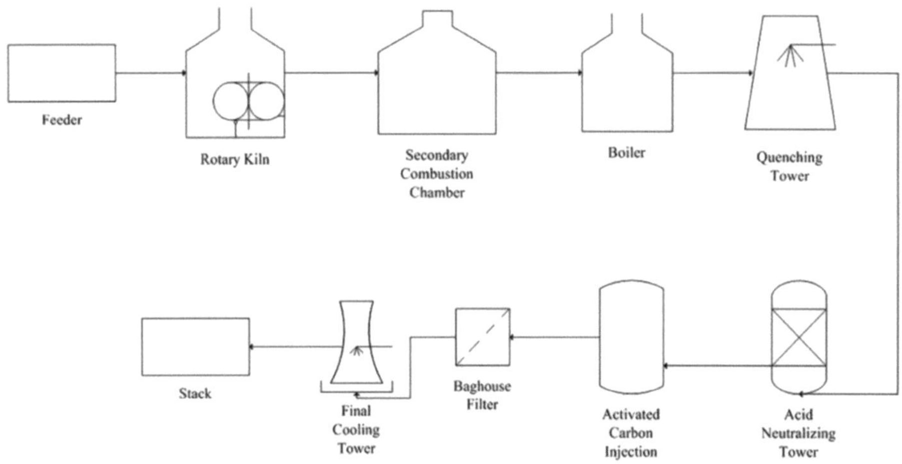

Figure 3. Flow diagram of sulfur dioxide circulation system [61].

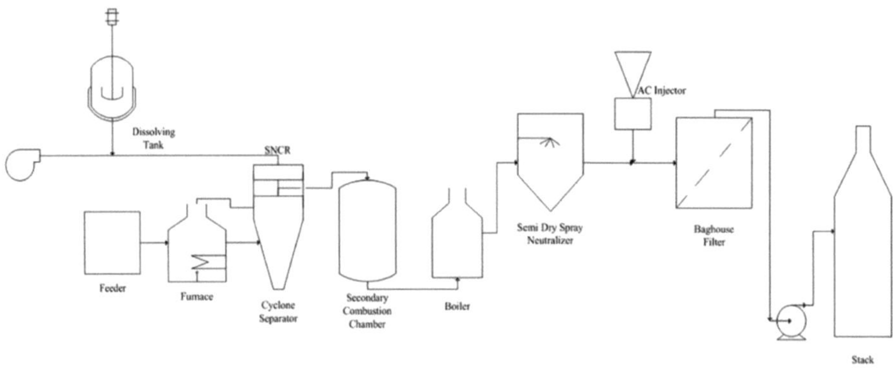

Figure 4. Flow diagram of the waste incinerator that uses thiourea (1: feeder; 2: furnace; 3: cyclone separator; 4: SNCR; 5: secondary combustion chamber; 6: boiler; 7: semi-dry spray neutralizer; 8: AC injector; 9: bag house filter; 10: stack; 11: dissolving tank) [60].

carbon dosing [61]. One important criterion to optimize the minimization of PCDD/PCDFs is the proper selection of sulfur compounds. In third technology, compounds of nitrogen such as ethanolamine, mono-ethanolamine (MEA), urea, ammonia, tri-ethanolamine, di-methylamine were added to control the formation of dioxins and furans. It was studied that thiourea is a suitable dioxin inhibitor with high S and N-content **Figure 4** [61].

7. Conclusions

The concentration of PCDD/PCDFs in environment has been increased up to the range of toxicity so these compounds may affect the biological systems. There are various means of production of PCDD/PCDFs which has significant impact on rise of levels of these compounds in ambient air. The concentrations of PCDD/PCDFs in the environment are not well controlled due to many reasons like temperature variations and seasonal pattern. Although a large no of research studies have been conducted to investigate and control PCDD/PCDFs but even then information is limited to get about these compounds therefore it is difficult to get rid of them. Further investigations should be conducted to establish a comprehensive approach to investigate the recent profiles of PCDD/PCDFs found in various parts of the world especially in developing countries. This chapter will help the stake holders in decision making process to establish a sustainable waste management system in future. New and better innovative research plans are necessary for addressing the problems related to pollution.

Author details

Muhammad Zubair* and Amina Adrees

*Address all correspondence to: muhammad.zubair@uog.edu.pk

Department of Chemistry, Faculty of Science, University of Gujrat, Pakistan

References

[1] Colbeck I, Nasir ZA, Ali Z. The state of ambient air quality in Pakistan—A review. Environmental Science and Pollution Research. 2010;**17**(1):49-63

[2] Zhang Q, Jiang X, Tong D, Davis SJ, Zhao H, Geng G, et al. Transboundary health impacts of transported global air pollution and international trade. Nature. 2017;**543**(7647):705

[3] Kumar JM, Deepika D, Srinithya B, Kalaichelvan P. Polychlorinated dibenzo P dioxins and furans—A review. International Journal of Current Research and Review. 2013;**5**(3):14

[4] Boffetta P, Gouas DA, da Costa AN, Abedi-Ardekani B, Hainaut P. Cancers of the Intestine, Liver, and Biliary Tract. Occupational Cancers. Springer; 2014. pp. 127-137

[5] Safe S, Hutzinger O, Hill TA. Polychlorinated Dibenzo-p-Dioxins and-Furans (PCDDs/PCDFs): Sources and Environmental Impact, Epidemiology, Mechanisms of Action, Health Risks. Springer Science & Business Media; 2012

[6] Hoogenboom RL, ten Dam G, van Bruggen M, Jeurissen SM, van Leeuwen SP, Theelen RM, et al. Polychlorinated dibenzo-p-dioxins and dibenzofurans (PCDD/Fs) and biphenyls (PCBs) in home-produced eggs. Chemosphere. 2016;**150**:311-319

[7] Urban JD, Wikoff DS, Bunch AT, Harris MA, Haws LC. A review of background dioxin concentrations in urban/suburban and rural soils across the United States: Implications for site assessments and the establishment of soil cleanup levels. Science of the Total Environment. 2014;**466**:586-597

[8] Alcock RE, Jones KC. Dioxins in the environment: A review of trend data. Environmental Science & Technology. 1996;**30**(11):3133-3143

[9] Cleverly D. The congener profiles of anthropogenic sources of chlorinated dibenzo-p-dioxins and chlorinated dibenzofurans in the United States. Organohalogen Compounds. 1997;**32**:430-435

[10] Boer F, Neuman M, Van Remoortere F, North P, Rinn H. X-Ray Diffraction Studies of Chlorinated Dibenzo-p-Dioxins. ACS Publications; 1973

[11] Hites RA. Dioxins: An Overview and History. ACS Publications; 2010

[12] Schwetz B, Norris J, Sparschu G, Rowe V, Gehring P, Emerson J, et al. Toxicology of Chlorinated Dibenzo-p-Dioxins. ACS Publications; 1973

[13] Van den Berg M, Birnbaum LS, Denison M, De Vito M, Farland W, Feeley M, et al. The 2005 World Health Organization reevaluation of human and mammalian toxic equivalency factors for dioxins and dioxin-like compounds. Toxicological Sciences. 2006;**93**(2):223-241

[14] Kulkarni PS, Crespo JG, Afonso CA. Dioxins sources and current remediation technologies—A review. Environment International. 2008;**34**(1):139-153

[15] Bumb R, Crummett W, Cutie S, Gledhill J, Hummel R, Kagel R, et al. Trace chemistries of fire: A source of chlorinated dioxins. Science. 1980:385-390

[16] Rawls RL, Swanson D, Meselson M. Dow finds support, doubt for dioxin ideals. Chemical and Engineering News (United States). 1979;**57**(7)

[17] Mocarelli P. Seveso: A teaching story. Chemosphere. 2001;**43**(4-7):391-402

[18] Stone R. Agent Orange's Bitter Harvest. American Association for the Advancement of Science; 2007

[19] Päpke O. PCDD/PCDF: Human background data for Germany, a 10-year experience. Environmental Health Perspectives. 1998;**106**(Suppl 2):723

[20] Huang H, Buskens A. Comparison of dioxin formation levels in laboratory gas-phase flow reactors with those calculated using the Shaub-Tsang mechanism. Chemosphere. 1999;**38**(7):1595-1602

[21] Sidhu S, Edwards P. Role of phenoxy radicals in PCDD/F formation. International Journal of Chemical Kinetics. 2002;**34**(9):531-541

[22] Addink R, Olie K. Mechanisms of formation and destruction of polychlorinated dibenzo-p-dioxins and dibenzofurans in heterogeneous systems. Environmental Science & Technology. 1995;**29**(6):1425-1435

[23] Delvaux T, Buekens P, Godin I, Boutsen M. Barriers to prenatal care in Europe. American Journal of Preventive Medicine. 2001;**21**(1):52-59

[24] Tuppurainen K, Halonen I, Ruokojärvi P, Tarhanen J, Ruuskanen J. Formation of PCDDs and PCDFs in municipal waste incineration and its inhibition mechanisms: A review. Chemosphere. 1998;**36**(7):1493-1511

[25] Ligon W, Dorn SB, May RJ, Allison MJ. Chlorodibenzofuran and chlorodibenzo-p-dioxin levels in Chilean mummies dated to about 2800 years before the present. Environmental Science & Technology. 1989;**23**(10):1286-1290

[26] Czuczwa JM, Hites RA. Environmental fate of combustion-generated polychlorinated dioxins and furans. Environmental Science & Technology. 1984;**18**(6):444-450

[27] Altwicker E. Relative rates of formation of polychlorinated dioxins and furans from precursor and de novo reactions. Chemosphere. 1996;**33**(10):1897-1904

[28] Abad E, Adrados M, Caixach J, Rivera J. Dioxin abatement strategies and mass balance at a municipal waste management plant. Environmental Science & Technology. 2002; **36**(1):92-99

[29] Stanmore B, Clunies-Ross C. An empirical model for the de novo formation of PCDD/F in medical waste incinerators. Environmental Science & Technology. 2000;**34**(21):4538-4544

[30] Karademir A, Bakoglu M, Ayberk S. PCDD/F removal efficiencies of electrostatic precipitator and wet scrubbers in izaydas hazardous waste incinerator. Fresenius Environmental Bulletin. 2003;**12**(10):1228-1232

[31] Fullana A, Conesa JA, Font R, Sidhu S. Formation and destruction of chlorinated pollutants during sewage sludge incineration. Environmental Science & Technology. 2004; **38**(10):2953-2958

[32] Eduljee G. Waste disposal in cement kilns: A review of dioxin formation and control. Environmental & Waste Management. 1999;**2**(1):45-54

[33] Quaß U, Fermann MW, Bröker G. Steps towards a European dioxin emission inventory. Chemosphere. 2000;**40**(9-11):1125-1129

[34] Lavric ED, Konnov AA, De Ruyck J. Dioxin levels in wood combustion—A review. Biomass and Bioenergy. 2004;**26**(2):115-145

[35] Marklund S, Andersson R, Tysklind M, Rappe C, Egebäck K-E, Björkman E, et al. Emissions of PCDDs and PCDFs in gasoline and diesel fueled cars. Chemosphere. 1990; **20**(5):553-561

[36] Rappe C, Andersson R, Bergqvist P-A, Brohede C, Hansson M, Kjeller L-O, et al. Overview on environmental fate of chlorinated dioxins and dibenzofurans. Sources, levels and isomeric pattern in various matrices. Chemosphere. 1987;**16**(8-9):1603-1618

[37] Wang X, Ni Y, Zhang H, Zhang X, Chen J. Formation and emission of PCDD/Fs in Chinese non-wood pulp and paper mills. Environmental Science & Technology. 2012; **46**(21):12234-12240

[38] Anderson DR, Fisher R. Sources of dioxins in the United Kingdom: The steel industry and other sources. Chemosphere. 2002;**46**(3):371-381

[39] Cieplik MK, Carbonell JP, Muñoz C, Baker S, Krüger S, Liljelind P, et al. On dioxin formation in iron ore sintering. Environmental Science & Technology. 2003;**37**(15):3323-3331

[40] Shiomitsu T, Hirayama A, Iwasaki T, Akashi T, Fujisawa Y. Volatilization and decomposition of dioxin from fly ash. In: NKK Technical Report-Japanese Edition. 2001. pp. 8-11

[41] Anthony E, Jia L, Granatstein D. Dioxin and furan formation in FBC boilers. Environmental Science & Technology. 2001;**35**(14):3002-3007

[42] Petroff BK, Roby KF, Gao X, Son D-S, Williams S, Johnson D, et al. A review of mechanisms controlling ovulation with implications for the anovulatory effects of polychlorinated dibenzo-p-dioxins in rodents. Toxicology. 2001;**158**(3):91-107

[43] Gray Jr LE, Kelce WR. Latent effects of pesticides and toxic substances on sexual differentiation of rodents. Toxicology and Industrial Health. 1996;**12**(3-4):515-531

[44] Rier S, Foster WG. Environmental dioxins and endometriosis. In: Seminars in Reproductive Medicine. New York, NY, USA: Thieme Medical Publishers, Inc.; 2003

[45] Gray Jr LE. Xenoendocrine disrupters: Laboratory studies on male reproductive effects. Toxicology Letters. 1998;**102**:331-335

[46] Pop VJ, Kuijpens JL, van Baar AL, Verkerk G, van Son MM, de Vijlder JJ, et al. Low maternal free thyroxine concentrations during early pregnancy are associated with impaired psychomotor development in infancy. Clinical Endocrinology. 1999;**50**(2):149-155

[47] Vulsma T. Impact of exposure to maternal PCBs and dioxins on the neonate's thyroid hormone status. Epidemiology. 2000;**11**(3):239

[48] Kerkvliet NI. Immunological effects of chlorinated dibenzo-p-dioxins. Environmental Health Perspectives. 1995;**103**(Suppl 9):47

[49] Kakeyama M, Tohyama C. Developmental neurotoxicity of dioxin and its related compounds. Industrial Health. 2003;**41**(3):215-230

[50] Stohs SJ. Oxidative stress induced by 2, 3, 7, 8-tetrachlorodibenzo-p-dioxin (TCDD). Free Radical Biology and Medicine. 1990;**9**(1):79-90

[51] Hebert C, Harris M, Elwell M, Birnbaum L. Relative toxicity and tumor-promoting ability of 2, 3, 7, 8-tetrachlorodibenzo-p-dioxin (TCDD), 2, 3, 4, 7, 8-pentachlorodibenzofuran (PCDF), and 1, 2, 3, 4, 7, 8-hexachlorodibenzofuran (HCDF) in hairless mice. Toxicology and Applied Pharmacology. 1990;**102**(2):362-377

[52] Panteleyev AA, Thiel R, Wanner R, Zhang J, Roumak VS, Paus R, et al. 2, 3, 7, 8-tetrachlorodibenzo-p-dioxin (TCDD) affects keratin 1 and keratin 17 gene expression and differentially induces keratinization in hairless mouse skin. Journal of Investigative Dermatology. 1997;**108**(3):330-335

[53] Pohjanvirta R, Tuomisto J. Short-term toxicity of 2, 3, 7, 8-tetrachlorodibenzo-p-dioxin in laboratory animals: Effects, mechanisms, and animal models. Pharmacological Reviews. 1994;**46**(4):483-549

[54] Unkila M, Pohjanvirta R, Tuomisto J. Biochemical effects of 2, 3, 7, 8-tetrachlorodibenzo-p-dioxin (TCDD) and related compounds on the central nervous system. The International Journal of Biochemistry & Cell Biology. 1995;**27**(5):443-455

[55] Smith AG, Clothier B, Robinson S, Scullion MJ, Carthew P, Edwards R, et al. Interaction between iron metabolism and 2, 3, 7, 8-tetrachlorodibenzo-p-dioxin in mice with variants of the Ahr gene: A hepatic oxidative mechanism. Molecular Pharmacology. 1998; **53**(1):52-61

[56] Rier SE, Martin DC, Bowman RE, Dmowski WP, Becker JL. Endometriosis in rhesus monkeys (*Macaca mulatta*) following chronic exposure to 2, 3, 7, 8-tetrachlorodibenzo-p-dioxin. Toxicological Sciences. 1993;**21**(4):433-441

[57] Heilier J-F, Donnez J, Lison D. Organochlorines and endometriosis: A mini-review. Chemosphere. 2008;**71**(2):203-210

[58] Johnson KL, Cummings AM, Birnbaum LS. Promotion of endometriosis in mice by polychlorinated dibenzo-p-dioxins, dibenzofurans, and biphenyls. Environmental Health Perspectives. 1997;**105**(7):750

[59] Vermeulen I, Van Caneghem J, Vandecasteele C. Indication of PCDD/F formation through precursor condensation in a full-scale hazardous waste incinerator. Journal of Material Cycles and Waste Management. 2014;**16**(1):167-171

[60] Lin X, Yan M, Dai A, Zhan M, Fu J, Li X, et al. Simultaneous suppression of PCDD/F and NOx during municipal solid waste incineration. Chemosphere. 2015;**126**:60-66

[61] Liu X, Wang J, Wang X, Zhu T. Simultaneous removal of PCDD/Fs and NOx from the flue gas of a municipal solid waste incinerator with a pilot plant. Chemosphere. 2015; **133**:90-96

Chapter 8

Emissions of Inorganic Trace Pollutants from Coal Power Generation

Patricia Córdoba

Additional information is available at the end of the chapter

http://dx.doi.org/10.5772/intechopen.79918

Abstract

Coal-fired power plants are a major source of emissions for a number of air pollutants including SO_2, NOx, particulate matter (PM), HCl, HF, Hg, and so on. Hazardous air pollutants such as As, Be, Cd, Cr, Pb, Mn, Ni, Se, and other metals are integral components of fine PM that are also emitted directly from coal-fired power plants. The potential problem of coal-fired power plants associated to the emissions of air pollutants can be treated by flue gas desulphurisation (FGDs), as the sulphur (S) will be retained in the FGD-gypsum ($CaSO_4{\cdot}2H_2O$), and eventually be disposed. Further, FGD chemistry (alkaline sorbent) allows the capture of many pollutants other than sulphur, such as F, As, B, Cl, Se or Hg, in a gaseous form and/or as PM. In this chapter, the current position of coal power generation and the generation of inorganic trace pollutants derived from it are presented and discussed. The partitioning, speciation, and fate of inorganic trace pollutants during pulverised coal combustion (PCC)-FGD are also reviewed.

Keywords: particulate matter, hazardous air pollutants, flue gas desulphurisation systems (FGD), partitioning, speciation

1. Introduction

Coal plays an essential role in our global energy scheme for power generation as most of the world's coal production is consumed mainly to generate electricity. However, coal is currently a target especially for Europe, where political and social opposition to coal is mounting as efforts intensify to limit CO_2 emissions. In the United States, coal industry also declined in 2017, and its trend toward long-term structural decline is all but sure to persist in 2018 [1] while in China concerns about their air quality and related health issues caused demand for coal to fall from 2014 to 2016 [2]. As coal consumption declines in most of developed countries, including those in Europe and the US, the future of coal is increasingly dependent on developing countries.

Most European power generation producers have ruled out the construction of new coal plants, and utilisation at existing coal-fired power plants in Europe is decreasing as renewable energy sources expand [2]. The demand for coal in China, on the other hand, the primary driver of coal since the 2000s, has just about peaked and is probably to begin declining in the early 2020s [2]. Instead, Chinese government is promoting a shift from coal to less polluting sources, including renewables, in order to address the country's air quality. However, under this entire scenario, coal demand will either continue to grow modestly or remain stable as coal is the world's most abundant energy resource. There are 1,139,331 million tonnes of proven coal reserves worldwide, sufficient to meet 153 years of global production. In comparison, proven oil and natural gas reserves are equivalent to around 50 and 53 years, respectively, at 2016 production levels [3]. Therefore, despite the decline in coal production and consumption, coal is and will be a reliable source for power generation.

Coal-fired power plants are a major source of emissions for a number of air pollutants including SO_2, NOx, particulate matter (PM), HCl, HF, and Hg among and solid residues, mostly ashes [4]. The potential problem of coal-fired power plants associated to the emissions of air pollutants can be treated by flue gas desulphurisation (FGDs), as the S will be retained in the $CaSO_4.2H_2O$, and eventually be disposed. Further, FGD chemistry (alkaline sorbent) allows the capture of many pollutants other than sulphur, such as F, As, B, Cl, Se or Hg [4–6], in a gaseous form and/or as PM [4].

Commercial power generation wet FGD system uses calcite ($CaCO_3$) from limestone, slaked lime ($Ca(OH)_2$), or a mixture of $Ca(OH)_2$ and alkaline FA sorbents, which react with SO_2 to form a Ca-S compound [7]. Although the wet limestone based process is by far the most widely FGD employed because of its high desulphurisation performance, reliability, and efficiency [7], the use of slaked lime has also been proposed since lime shows over limestone higher pH values and better reactivity defined as capacity to provide alkalinity.

In this chapter, we explore the current position of coal power generation as a source of inorganic trace pollutants emissions. Firstly, the current situation of coal combustion for energy production and the generation of inorganic trace pollutants are presented and discussed. Secondly, the classification of trace elements according to their affinities in coal and their partitioning and behaviour during combustion is evaluated. Thirdly, the partitioning, speciation, and fate of inorganic during PCC-FGD are reviewed.

2. Power generation from coal

2.1. Coal

Coal is defined by the American Society for Testing and Materials [8] as 'brown to black combustible sedimentary rock composed principally of consolidated and chemically altered plant'. Over 90% of the coal consumed in Europe, the US, and other countries is used to generate electricity. Coal power is also used as a basic industry source for making steel, cement and paper, among other industries.

Coal-fired power plants currently fuel 40% of global electricity, and, in some countries, coal fuels a higher percentage of electricity [9]. Owing to the Paris climate agreement, in Europe, the use of coal for power generation retreated for the fifth successive year in 2017. The European Power Sector 2017 reported that coal's share of Europe's total power generation fell to 20% last year, while the share from renewables increased to 30%. Nevertheless, Europe's progress in reducing the use of carbon-intensive power is gradual and uneven.

In 2017, coal production in the US was higher than in 2016, which was attributed in part to the bankruptcy-caused restructuring of several major coal producers, which resulted in lower production costs [10]. China produced 3.45 billion tonnes of raw coal in 2017, a year-on-year rise of 3.2%, showed data released by the National Bureau of Statistics on January 18 [11].

However, the role of coal in power generation is set to continue. Coal currently fuels 40% of the world's electricity and is forecast to continue to supply a strategic share over the next three decades [12]. There are 1,139,331 million tonnes of proven coal reserves worldwide, sufficient to meet 153 years of global production (**Table 1**). Total proven coal reserves are shown for anthracite and bituminous (including brown coal) and sub-bituminous and lignite. In comparison, proven oil and natural gas reserves are equivalent to around 50 and 53 years, respectively, at end 2016 production levels [3].

Million tonnes	**Anthracite and bituminous**	**Sub-bituminous and lignite**	**Total**	**Share of total**	**R/P ratio**
Total North America	226,906	32,469	259,375	23%	356
Total S. & Cent. America	8943	5073	14,016	1.2%	138
Total Europe & Eurasia	153,283	168,841	322,124	28%	284
Total Middle East & Africa	14,354	66	14,420	1.3%	54
Total Asia Pacific	412,728	116,668	529,396	46%	102
Total World	816,214	323,117	1,139,331	100%	153

Table 1. Proven coal reserves worldwide.

2.2. Coal combustion

Pulverised coal combustion (PCC) is the most widely technology to produce energy from coal. In this process, coal is pulverised by coal mills in order to be ground to fineness (such that 70–80%). The oxidant, an air stream, is injected into the coal mills to form an air-coal mixture, which is carried by conveyors to feed individually each burner in the boiler. The secondary/extra air necessary for the combustion process is injected into the burners via secondary and tertiary air nozzles to initiate combustion, depending on the coal rank, at temperatures from 1300 to 1700°C [13]. The generated heat from the combustion process is handed to water tubes placed in the boiler walls, so as the water in it turns into a steam (34–44 bars and 540°C). This steam powers a high vapour pressure turbine (38–51 bars and 3000 rpm) and later is recirculated and expanded to power medium and low vapour pressure turbines. After the recirculation, the steam is delivered in a condenser where it is cooled and sent to the boiler to heat it again. Finally, the combustion gases are used to overheat the water from the economiser and the air used in combustion (**Figure 1**).

Coal combustion by-products (CCBs) are the materials produced from the combustion of coal for the production of electricity. CCBs primarily include: bottom ash (BA) and/or boiler slag (BS), fly ash (FA), and flue gas desulphurisation (FGD)-gypsum. The coarse fraction of ashes, BA or BS, is generally removed from the bottom of the boiler while the smallest fraction of these, FAs, can either be captured, depending on the depuration train technology, in an electrostatic precipitator (ESP) or fabric filter (FF) [5, 6]. ESPs and FFs operate with a very high efficiency (>99%); however, a small fraction of FAs may escape from the control and reach the FGD system (4) where S will be retained in the $CaSO_4.2H_2O$, and eventually be disposed. Further, FGD chemistry (alkaline sorbent) allows the capture of many pollutants other than S, such as F, As, B, Cl, Se or Hg [4–6], in a gaseous form and/or as PM (**Figure 1**).

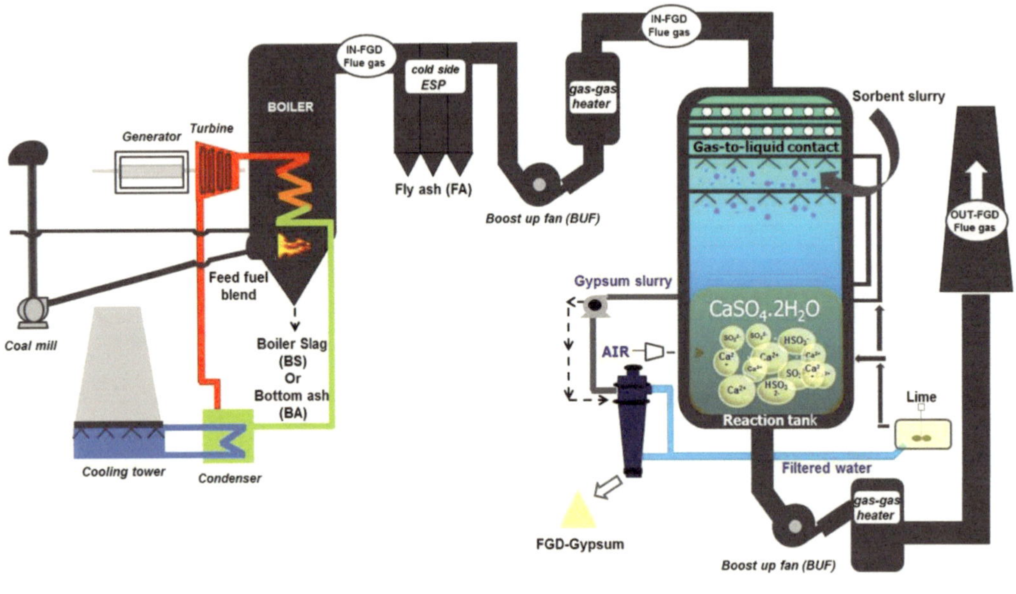

Figure 1. Configuration of a PCC plant equipped with an FGD system.

There are alternatives other than PCC for coal power generation. Fluidised bed combustion (FBC) is a technology in which, coal or any solid fuel is suspended on jets of air during the combustion process, which allows for a more efficient mixing of gas and solids and results in more efficient heat transfer. One primary advantage of FBC technology is that the combustion reaction can take place at lower temperatures, which reduces the formation of toxic nitrous oxide (N_2O). In addition, SO_2 can be more cheaply and easily removed during combustion than using FGD technology. Lastly, FBC can support the addition of other solid fuels such as biomass into the reaction mixture along with coal [14]. Currently, three different types of FBC are available, the circulating FBC (CFBC), the bubbling FBC (BFBC), and pressurised FBC [7] (**Table 2**).

http://dx.doi.org/10.5772/intechopen.79918

	Principle	Fluidisation	SO_2 retention	NOx emissions
Circulating FBC	Addition of limestone during combustion to react with SO_2	3–9 m/s	98%	120 mg/Nm^3
Bubbling FBC	Addition of limestone during combustion to react with SO_2	1–3 m/s		200–300 mg/Nm^3
Pressurised FBC	Air is pressurised using a gas turbine compressor		SOx level of approximately 5 ppm through in-bed desulphurisation	NOx level of approximately 100 ppm

FBC: fluidised bed combustion.

Table 2. FBC technologies.

2.3. Co-combustion

Co-combustion means simultaneous combustion of two or more fuels in the same plant for energy production. The current and future energy policy aims at increasing the share of renewable energy in worlds energy supply. One possibility to enhance energy production by renewable sources within a short term is co-combustion [15].

Co-combustion can be carried out in different ways according to the operating conditions of the existing power plant. Accordingly, co-combustion can be carried out by using (1) a small amount (a few percent of total fuel power) of co-combustion matter fired together with coal in a boiler, originally designed for coal; (2) a small amount of fuel with a high heating value fired together with a fuel having a low heating value such as sewage sludge that needs thermal support to attain a desired combustion temperature; and by (3) spontaneous use of co-combustion with fuels in any ratio, depending on price, availability and local supply conditions [16].

The most common co-combustion matters fired together with coal can be petroleum coke, sewage sludge, and biomass. Petroleum coke is a carbon-rich solid that emerges as a by-product of crude oil refining and other 'cracking' processes, in which complex organic molecules such as heavy hydrocarbons are broken down into simpler and more valuable lighter petroleum products. Depending on the process of production, there are three types of petroleum coke: delayed coking (93% of world production), fluid coke (6% of pet-coke production), and flexi-coking (1% of pet-coke production) [17]. These three types of pet-coke have higher calorific values than coal and contain less volatile matter and ash (less than 1–2%); qualities that make petroleum coke an attractive alternative for power producers.

Sewage sludge, a type of bio-solid, is a by-product of wastewater treatment. In the context of new energy policies and legislation, the use of sewage sludge for power generation is promoted, not only by environmental considerations (sewage sludge is burned with zero net CO_2 emissions), but also by economic considerations due to the gain of reducing disposal charges [18].

Biomass represents a lot of various materials, either waste materials or special energetic plants. Fuels based on wood biomass (sawdust, shavings, chips, tree-bark) can also be used for the production of high-quality biofuels, such as wooden briquettes and pellets, or can be co-combusted with coal [19]. Although biomass co-firing is one of the most effective means of reducing Green House Gas (GHG) emissions in PCC power plants [16], the chemical properties of biomass have implications for combustion quality and/or impacts on equipment when it is fed as co-matter [20].

2.4. Oxy-combustion

Oxyfuel combustion is one of the leading technologies considered for capturing CO_2 from power plants with carbon capture and storage (CCS) [21]. This involves the process of burning the fuel with nearly pure oxygen instead of air, which results in higher flame temperatures. For this reason, the mixture is diluted with a portion of the resulting flue gas composed primarily of CO_2.

The oxygen for oxy-fuel combustion is provided by two oxidiser streams: primary and secondary stream (**Figure 2**). The primary oxidiser stream supplies part of the oxygen necessary for combustion by carrying the pulverised coal into the wall firing burners placed in the oxy-boiler while the secondary oxidiser streams consists of oxygen that comes from the air separation unit (ASU) mixed with a portion of the recycled flue gas (FGR) to moderate flame temperatures and obtain a boiler heat transfer profile similar to that of air-firing in the case of retrofit. Depending on oxy-fuel combustion configuration, the secondary stream will constitute the equivalent to the secondary, tertiary, and overfire (if necessary) air flows (OFA) when oxygen has been added. Several options exist for the position where the secondary recycle stream is taken. As an example, at the largest oxy-PCC demonstration plant [22], the CO_2-rich FGR is divided into three oxidant streams: CB1, CB2, and CB3 (**Figure 2**), and it can also operate with an OFA system as part of an overall NOx reduction strategy by staging the combustion process. The OFA system normally reduces the oxygen availability early in the oxy-combustion process by reintroducing it later through ports located above the combustion zone (**Figure 2**).

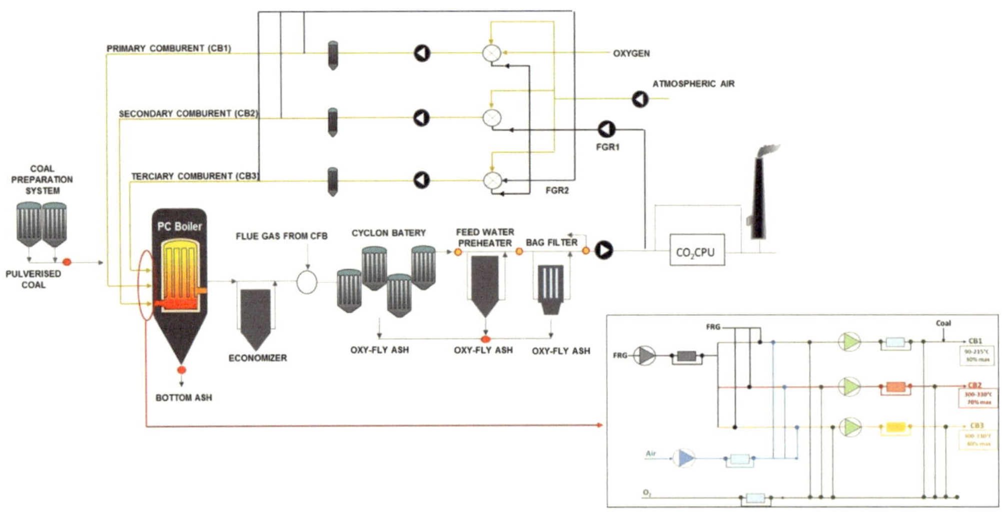

Figure 2. Configuration of the largest oxy-fuel demonstration plant [13].

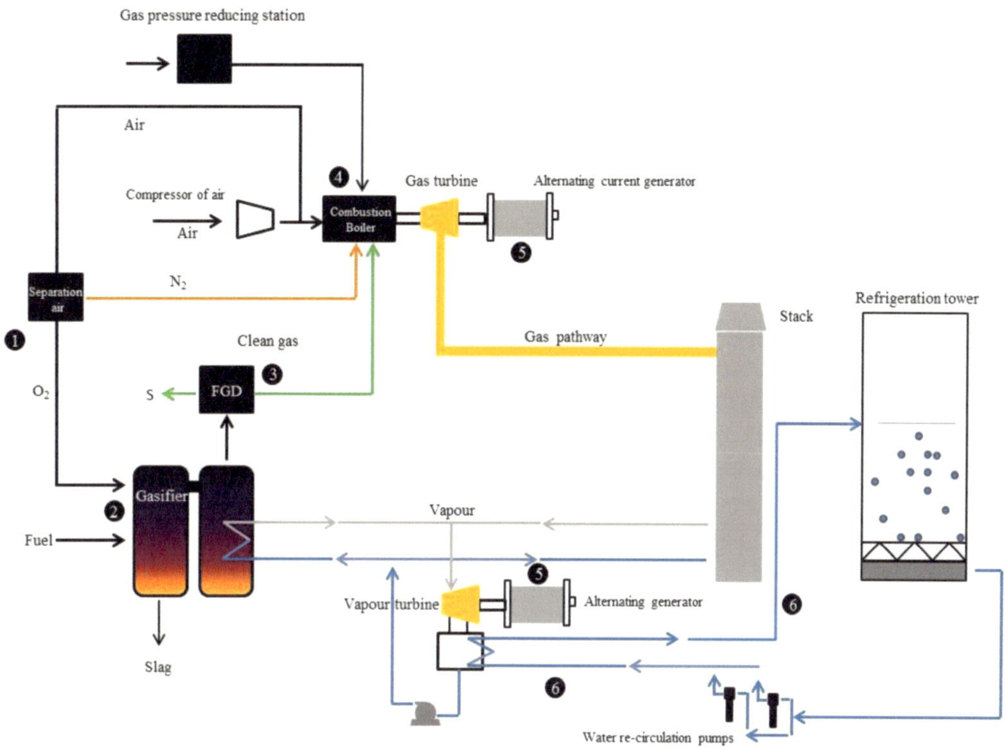

Figure 3. Configuration of a integrated gasification combined cycle (IGCC) power plant [3].

2.5. Gasification of coal

While the goal of coal combustion is to produce the maximum amount of heat possible by oxidising all the combustible material, coal gasification is a thermochemical process in which coal is converted into CO, H_2 and CH_4 synthesis gas (syngas) by means of the partial combustion with air or water vapour and low levels of O_2 [23]. Different technologies of coal gasification that are currently available are shown in **Table 3** [23]:

	Principle	Temperature (°C)	Pressure (bar)
Fixed bed gasifier	Coal moves in counter current to the gas flow	400–1100	10–100
Fluidised bed gasifier	Gaseous agents are blown through a flow of solid particles to keep these in a state of suspension	800–1050	10–25
Entrained bed gasifier	Coal and gasifier agents flow in the same direction with velocities higher than the rest of gasifier	1200–1600	25–40
Integrated gasification in combined cycle (IGCC)	Use of the CO-and H_2-rich gaseous in a gaseous turbine and the use of the leftover heat in a vapour turbine	~1500	25

Table 3. Coal gasification technologies.

Among the current technologies of coal gasification, integrated gasification combined cycle (IGCC) is the most common (**Figure 3**). The IGCC is characterised by the use of the CO-and H_2-rich gaseous stream resulting from the gasification, usually of coal and/or petroleum coke, in a gaseous turbine and the use of the leftover heat in a vapour turbine (**Figure 3**). The IGCC entails an initial fractionation of air (1) in the ASU. In this unit, cryogenic distillation separates O_2 from air by liquefying air at very low temperatures (−300°F). Ambient air is compressed in multiple stages with inter-stage cooling then further cooled with chilled water. Residual water vapour, carbon dioxide and atmospheric contaminants are removed in molecular sieve adsorbers. In the gasifier, (2) coal reacts with the O_2 and H_2O stream to generate a syngas at high temperature (~1500°C, 25 bar) [24]. Heat in the gasifier liquefies the coal ashes and subsequently the molten ash is quenched and crushed at the bottom of the gasifiers before being dewatered for disposal. The syngas stream passes to the flue gas depuration train (3) in which SO_2 and other gaseous pollutants are removed from the gaseous stream. The cleaned gas is then burned in a combined cycle power generation unit (4). In this unit, the gaseous stream is expanded in a gaseous turbine, whereas the leftover heat is expanded in the vapour turbine both connected to generators of electric energy (5). The water vapour condensates as a consequence of the heat exchange with water from the refrigeration tower (6).

The formation of NOx in the combustion chamber of the gas turbine is suppressed by saturation of the fuel gas with steam prior to combustion and by dilution with N_2 from the air separation unit.

3. Behaviour of trace elements during coal combustion

Trace elements are introduced in PCC from coal or co-combustion material. According to their different contents, these elements can be divided into (1) major elements (C, H, O, N, S) whose content is >1000 ppm; (2) minor elements which include coal mineral matters (Si, Al, Ca, Mg, K, Na, Fe, Mn, Ti) and halogens (F, Cl, Br, I), with concentrations between 100 and 1000 ppm; and (3) trace elements with concentration < 100 ppm. It is generally accepted that trace elements combination and contents differ from one coal to another due to the different coalification processes [25–27]. The correlation between the organic content or the different mineral phases in coal with the content of major, minor, and trace elements allows the establishment of trace elements affinities in coal as follows [28]:

1. Elements with inorganic affinity:
   - *Clays and feldspars* (Al, K, Mg, Na, P, Ti, Li, Cr, Ni, Cu, Ga, Rb, V, Sr., Y, Sn, Cs, Ba, Ta, Pb, Bi, Th, U, Zr, and REEs).
   - *Sulphide minerals* (S, Fe, Ni, Co, Cu, Zn, As, Se, Mo, Cd, Sb, Hg, W, Pb, and Tl).
   - *Carbonate minerals* (C, Ca, Mn, and Co).
   - *Sulphate species* (S, Ca, Fe, Ba).

- ***Heavy minerals*** (B, Ti, Th, and Zr).
- ***Several mineral phases*** (Co and W) in carbonates and sulphides; Ni, Cu, and Pb (clays and sulphides).

2. Elements with total or partial organic affinity (C, N, S, Be, B, Ge, V, and W).

Studies on the fate of trace elements during combustion have shown that their volatility depends on their affinities and on the physical changes and chemical reactions of these elements with S or other volatile elements during combustion [29–31]. **Figure 4** shows the behaviour and fate of elements in accordance with their volatile behaviour at the boiler and at the ESP, respectively, during combustion according to Córdoba et al. [32, 33]. This classification of the volatile behaviour of trace elements during combustion is in agreement with most of the literature [34–36], except for some specific elements, where elements are classified into three groups (**Figure 4**).

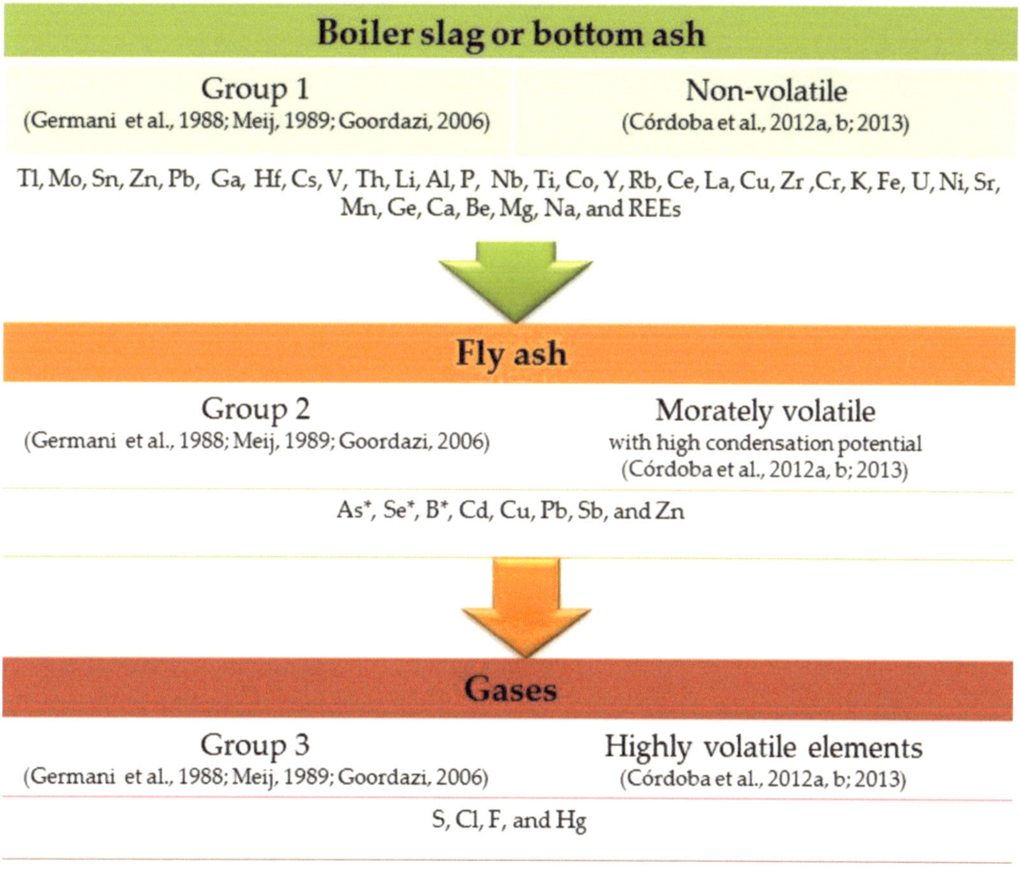

Figure 4. Behaviour of trace elements during coal combustion.

Elements classified as non-volatile and moderately volatile with condensation potential, respectively, from Córdoba et al. [32, 33] are in line with Group 1 and 2 elements from the abovementioned literature. However, elements such as As, Se and especially B can also be classified as highly volatile, which would correspond to Group 3.

The partitioning and fate of trace elements during combustion, discussed above, may be different by the use of secondary fuels. As discussed in Section 2.2, the use of co-matters such as petroleum coke, sewage sludge and/or biomass may have implications for combustion quality and/or modify the chemical environment of gaseous pollutants. Co-firing petroleum coke, for instance, may modify the chemical environment of Cl and S because of the resultant high concentrations of HCl and SOx, respectively, in the flue gas. An increase in the HCl concentration favours the formation of gaseous species, whereas increasing concentration of SO_2 in the gas composition enhances the formation of sulphate condensed species [37]. In addition, the heavy metals contents of the ash are generally high with Vanadium (V) and Nickel (Ni) contents ranging from 500 to 3000 ppm, although pet-cokes with >10,000 ppm V can also be found [38]. Molybdenum (Mo) can also be present in relatively high concentrations in petroleum cokes. The organic affinity of Mo, V and Ni in petroleum coke favours their volatility during pulverised coal combustion (PCC) and later condensation on the finest particles of FAs.

The main drawback of the sewage sludge combustion, on the other hand, is mostly related to high NOx emissions. The level of some toxic heavy metals and Cl in the raw material may also increase the emissions of hazardous pollutants (metals and dioxins).

3.1. Elements concentrated in coal combustion solid residues

There are some elements that either tend to get concentrated on the coarse residues BS or BA, partition equally between BS or BA and FA particulates, or to get enriched on the fine-grained particles, PM, which may escape particulate control systems.

BA is a granular material removed from the bottom of dry boilers, which is much coarser than FA though also formed during the combustion of coal. BS, on the other hand, is a vitreous grained material deriving from coal combustion in boilers at temperatures of 1500–1700°C, followed by wet ash removal of wet bottom furnaces [39].

FA is a fine powder made up of spherical high vitreous particles with Fe-oxides and Al-Si species, and irregular unburned coal and ash particles. The contents of principal oxides are usually in a descending order: $SiO_2 > Al_2O_3 > Fe_2O_3 > CaO > MgO > K_2O$. Fly ash also contains many trace elements, some of which are of environmental concern. Commonly, elements such as Cr, Pb, Ni, Ba, Sr., V and Zn are present in significant quantities. Coal aluminous-silicate impurities, mainly clays, with much lower proportion of feldspars, melt during combustion and rapidly shape themselves into spherical droplets [40]. The chemical composition of FAs may differ depending on the technology of combustion but especially on the characteristics of the feed coal. While coal combustion FA is constituted by an aluminous-silicate glass, with Ca, Fe, Na, K, Ti, and Mn impurities, and variable amounts of quartz, mullite, lime, haematite, magnetite, gypsum and feldspars, IGCC FA is characterised by a predominant Al-Si glass matrix and variety of fine crystalline reduced species (mostly metal sulphides) as a consequence of the low pO_2 at which fuel is burned [41].

3.2. Elements emitted in the flue gas

As aforementioned, most of the trace metals (As, Cd, Cr, Cu, Hg, Ni, Pb, Se, Sb, Tl, V, and Zn) may be released during combustion, emitted with a different form of occurrence (e.g. from sulphide in coal to oxides and chlorides in flue gas), and/or condense onto the surface of smaller particles in flue-gas streams. Therefore, most of trace metals are retained in particulate control devices and only specific high volatile metals may escape from ESP and reach FGD systems in a gaseous mode of occurrence. In this regard, FGD chemistry also allows the capture of many pollutants other than S, such as F, As, B, Cl, Se or Hg [41–46] both in a gaseous form and/or as PM. Thus, importantly, FGD systems can also be considered as a measure for the PM abatement emissions.

The abatement of NOx (NO, NO_2, and N_2O) emissions is based on the De-nitrification (DeNOx) process that aims at reducing NOx into N_2 and H_2O. The emission control systems that are in use to carry out the DeNOx process are: selective catalytic reduction (SCR) and selective non-catalytic reduction (SNCR).

4. Chemistry, partitioning and fate of inorganic trace pollutants during PCC-FGD

According to the foregoing discussion, trace elements during combustion may get concentrated on the coarse residues BS or BA, partition equally between BS or BA and FA particulates, emitted with a different form of occurrence (e.g. from sulphide in coal to oxides and chlorides in flue gas), and/or condense onto the surface of smaller particles in flue-gas streams. Either way, most of trace metals are retained in particulate control devices and only specific high volatile metals may escape from ESP and reach FGD systems in a gaseous mode of occurrence.

In the FGD systems, under operational conditions of water re-circulation, inorganic trace pollutants in FGD waters may reach equilibrium and a subsequent saturation in the water stream after a number of water re-circulations in the scrubber. The gradual increase in the concentration of inorganic trace pollutants from the sub-saturation to equilibrium and/or saturation because of continuous water re-circulation in the scrubber, accounts for enriched inorganic trace pollutants in the re-circulated water. Other elements retained in high proportions by gypsum sludge and/or FGD-gypsum do not pose this problem because they are extracted from the system by the gypsum by-product that is used for different applications or for landfilling [4]. The general trends of the inorganic trace pollutants in through the PCC to FGD are reported below.

Arsenic is present as As-sulphides species in raw coals and it is mostly released as $As_2O_{3\,(g)}$ [47] during PCC. In the boiler, gaseous As_2O_3 can be chemisorbed on the FA surface and/or remain in the gas phase. The chemisorption of As_2O_3 on FAs, which will depend on the temperature and gas composition, may occur via reaction with CaO to form Ca_3AsO_4. The small fraction of FAs in the flue gas that escapes from the control would be the main route by which Ca_3AsO_4-FA would enter the FGD. Accordingly, Ca_3AsO_4-FA would pass through the sprayers and dissolve to AsO_4^{3-} in the aqueous phase of the sorbent slurry. If As_2O_3 remains in the gas phase during the post-combustion atmosphere, a proportion of gaseous As_2O_3 could enter the FGD either by reacting with moisture in the flue gas to form H_3AsO_4 and condense in the scrubber as the flue gas undergoes a rapid quench (50–60°C), or by diffusing through the gas to the aqueous phase of the sorbent slurry where As_2O_3 would get hydrated to H_3AsO_4. Depending on the operating FGD conditions, As partitioning and fate may differ. Generally, As is mostly partitioned in the FGD-gypsum with values comprised in the range 90–99%.

Boron is largely organically associated in coals although a fraction can also be associated to aluminium silicates. Boron is generally released as H_3BO_3 and HBO_2 [48] during PCC. Since the chemisorption mechanism of H_3BO_3 and HBO_2 on FAs has not been documented, the main route by which H_3BO_3 can reach the FGD is with the incoming FGD flue gas. In the FGD gas-to-liquid contact zone, H_3BO_3 may diffuse through the gas to the aqueous phase of the sorbent slurry when flue gas passes through the sprayers and remain in the FGD reaction tank. Depending on the operating FGD conditions, B partitioning and fate may differ but in general, B is removed in the aqueous effluent (filtered water) and only a fraction of B may remain in the flue gases. However, since B concentration in FAs is relatively high and largely leachable, we can assume that FA containing B species such as $CaHBO_3$ can also contribute to increase B content in the in the FGD-gypsum [49].

Selenium casually present as selenide in coal is volatilised as elemental Se^0 and SeO_2. In the boiler, gaseous SeO_2 can be chemisorbed on the FA surface to form the stable $CaSeO_3$. The main route by which Se would reach the FGD is the small fraction of FAs escapes from the control and reaches the FGD. In such a case, Se chemisorbed in FAs could dissolve in the aqueous phase of the lime slurry to form an array of aqueous Se-complexes such as selenosulphate ($SeSO_3^{2-}$), selenotrithionate ($Se(SO_3)_2^{2-}$) and selenopentathionate ($Se(S_2O_3)_2^{2-}$) by reaction with polyoxysulphur donors [50]. Selenium can either be removed in the aqueous effluent (filtered water) or partitioned in the FGD-gypsum. However, a small fraction of Se can remain in the flue gases and be emitted into the atmosphere.

Mercury occurs in coals in mineral sulphide impurities, although other forms of occurrence, such as Hg-Se species, have been described [51]. During combustion, Hg is released as elemental Hg (Hg^0). During post-combustion, and with decreasing temperature, $Hg^0_{(g)}$ may remain as a monatomic species or may oxidise to Hg_2^{2+} and Hg^{2+} compounds. The reaction of $Hg^0_{(g)}$ with $HCl_{(g)}$ or $Cl_{2(g)}$ to form $HgCl_{2(g)}$ is generally considered to be the dominant Hg transformation mechanism in coal combustion flue gas [51]. The main route by which gaseous Hg can reach the FGD is with the incoming FGD flue gas. Gaseous compounds of Hg^{2+} are generally water-soluble and can dissolve in the aqueous phase of the sorbent slurry of wet FGD systems [52]. However, gaseous Hg^0 is insoluble in water and therefore does not dissolve. It is speculated that some of the absorbed Hg ($HgCl_2$) can be converted back to Hg^0 and re-emitted, S (IV) being the main precursor of Hg^0 re-emission [53–55].

Coal *chlorine* is released primarily as HCl during combustion in the high temperature zone of a boiler [56]; as the combustion gases cool (430–475°C), a proportion of HCl can partially be oxidised to Cl_2. The main route by which HCl can reach the FGD is with the incoming FGD flue gas. In the FGD gas-to-liquid contact zone, HCl may diffuse through the gas to the aqueous phase of the sorbent slurry when flue gas passes through the sprayers and react with cations such as Ca^{2+}, Na^{+}, Mg^{2+}, etc. to form highly soluble salts. Generally, Cl in FGD systems is removed in the aqueous effluent (filtered water) and only a fraction may remain in the flue gases.

Fluorine is released as HF during combustion. The main route by which gaseous HF can reach the FGD is with the incoming FGD flue gas. Depending on the operating FGD conditions, F partitioning and fate may differ. HF contained in the flue gas may be dissolved in the aqueous phase of the sprayed droplets of limestone slurry, if a limestone-based FGD, giving rise to the formation of CaF_2. However, in the presence of aluminium compounds, HF may also react with limestone to form Al-F compounds, typically represented by $CaAlF_3(OH)_2$-CaF_2 [57]. These compounds can coat the surfaces of the limestone particles and consequently cause a decrease the reactivity of limestone [58, 59]. In addition, if F is partitioned in the FGD-gypsum, it can play a crucial role in the leaching potential of the FGD gypsum end-product as a consequence of the precipitation of F solid species on FGD-gypsum surface. In order to avoid this, the use of additives has been proposed as measure for the optimisation of the SO_2 removal efficiency and for reducing the precipitation of F solid species, respectively.

Heavy metals such Zn, Cu, Cr, Ni, Mn and U tend to form highly soluble aqueous complexes with SO_4^{2-}. Depending on the operating FGD conditions, their partitioning and fate may differ. Acidic conditions, for example, contribute to the stabilisation of metals in solution, therefore, in the aqueous effluent. As a result, very low amounts of heavy metals are found in the leachates of the FGD-gypsum. Alkaline conditions, however, contribute to the precipitation of these elements in the FGD-gypsum.

5. Conclusions

Coal is currently a target to accomplish with the Paris climate agreement for both countries and companies. As a consequence, in 2016, world coal production fell by 6.2%, the largest decline on record. However, coal is the world's most abundant energy resource, meaning that despite the decline in coal production and consumption, coal is and will be a reliable source for power generation. The most negative consequence of coal combustion is the emissions of a number of air pollutants including SO_2, NOx, PM, HCl, HF, Hg and As, Be, Cd, Cr, Pb, Mn, Ni, and Se, integral components of fine PM. A number of studies have shown that FGD chemistry allows also the capture of many pollutants other than S, such as F, As, B, Cl, Se or Hg both in a gaseous form and/or as PM. Most of specialised literature reports that most of trace elements in FGD systems are removed in the aqueous effluent (filtered water) and only a fraction of a few remain in the flue gases (such as B, Hg, and Se). According to these studies, it can be concluded that wet limestone FGD systems reach high retention efficiencies for trace elements (>90%). However, it is also important to note that the retention efficiency of FGD system for trace elements may be reduced because of the emission of fly dust in evaporate droplets saturated with gypsum. In view of the discussion exposed in the review, we can conclude that the emissions of inorganic trace pollutants from coal power generation depend on each facility including the flue gas depuration train as well as the operating conditions associated to it.

Acknowledgements

The author gratefully acknowledges the Institute of Environmental Assessment and Water Research, Spanish National Research Council (IDÆA-CSIC).

Author details

Patricia Córdoba

Address all correspondence to: patricia.cordoba@idaea.csic.es

Institute of Environmental Assessment and Water Research, Spanish National Research Council (IDÆA-CSIC), Barcelona, Spain

References

[1] Why coal will keep burning. [Internet]. 2018. Available from: https://www.google.es/search?q=why+coal+will+keep+burning&oq=why+coal+will+&aqs=chrome.2.69i57j69i60j0l4.10408j0j8&sourceid=chrome&ie=UTF-8 [Accessed: 2018-06-17]

[2] IEEFA Outlook. [Internet]. 2018. Available from: http://ieefa.org/wp-content/uploads/2018/01/U.S.-Coal-More-Market-Erosion-_-February-2018.pdf [Accessed: 2018-06-17]

[3] British Petroleum (BP). Statistical Review of World Energy [Internet]. 2017. Available from: https://www.bp.com/content/dam/bp/en/corporate/pdf/energy-economics/statistical-review-2017/bp-statistical-review-of-world-energy-2017-full-report.pdf [Accessed: 2018-05-02]

[4] Córdoba P. Partitioning and speciation of trace elements at two coal-fired power plants equipped with a wet limestone flue gas desulphurization (FGD) system [thesis]. Barcelona: Universitat Politènica de Catalunya; 2013

[5] Clarke L. The fate of trace elements during combustion and gasification: An overview. Fuel. 1993;**72**(6):731-736

[6] Meij R. Trace element behaviour in coal-fired power plants. Fuel Processing Technology. 1994;**39**:199-217

[7] Reference Document on Best Available Techniques for Large Combustion Plants (BREFs). [Internet]. Available from: http://eippcb.jrc.ec.europa.eu/reference/BREF/LCP/JRC107769_LCP_bref2017.pdf [Accessed: 2018-05-010]

[8] American Society for Testing and Materials (ASTM). Annual book of ASTM standards. Section 5: Petroleum Products, Lubricants, and Fossil Fuels. 5.05: Gaseous Fuels: Coal and Coke; 1991

[9] World Coal Association. [Internet]. Available from: https://www.worldcoal.org/coal/uses-coal/coal-electricity [Accessed: 2018-06-22]

[10] US Energy Information Association. [Internet]. Available from: https://www.eia.gov/todayinenergy/detail.php?id=34992 [Accessed: 2018-06-22]

[11] Sxcoal. [Internet]. Available from: http://www.sxcoal.com/news/4567443/info/en

[12] World Energy Council. [Internet]. Available from: www.worldenergy.org/data/resources/resource/coal/ [Accessed: 2018-06-17]

[13] International Energy Agency (IEA). Clean Coal Centre. [Internet]. 2010. Available from: http://www.iea-coal.org.uk/site/2010/database-section/ccts/pulverised-coal-combustion-pcc [Accessed: 2018-06-17]

[14] Ground Truth Trekking. [Internet]. Available from: http://www.groundtruthtrekking.org/Issues/AlaskaCoal/CoalCombustionMethods.html [Accessed: 2018-06-21]

[15] TuDelft. [Internet]. Available from: https://repository.tudelft.nl/islandora/object/uuid%3Ac2b16256-45f1-419c-854a-7b24ad29d770

[16] Co-combustion a summary of technology. [Internet]. 2007. Available from: http://www.energy-pathways.org/pdf/R5_co-combustion.pdf

[17] Koffi M, da Silva RJ. Control of the SOx emissions during the petroleum coke combustion in fluidized bed. In: 19th International Congress of Mechanical Engineering; November 5-9, 2007; Brasília, DF

[18] Cenni R, Frandsen F, Gerhardt T, Spliethoff H, Hein KRG. Study on trace metal partitioning in pulverized combustion of bituminous coal and dry sewage sludge. Waste Management. 1998;**18**:433-444

[19] Bartoňová L, Klika Z, Milčák P. Determination of organic carbon content in ash from fluidised-bed coal combustion. Odpadové Fórum. 2008. pp. 22-24

[20] Leckner B. Co-combustion—A summary of technology. Thermal Science. 2007;**11**(4):5-40

[21] Stangera R, Walla T, Spörl R, Paneru M, Grathwohl S, Weidmann M, Scheffknecht G, McDonald D, et al. Oxyfuel combustion for CO_2 capture in power plants. International Journal of Greenhouse Gas Control. 2015;**40**:55-125

[22] Córdoba P, Diego R. Mass balance and partitioning of trace elements under oxy-coal combustion: First experiences. Fuel. 2018;**215**:892-903

[23] European Commission (EC). Gasification Clean Coal Technologies Hand Book. [Internet]. 2000. Available from: https://ec.europa.eu/energy/sites/ener/files/documents/2003_cofiring_eu_bionet.pdf [Accessed: 2018-03-13]

[24] Gasificación Integrada en Ciclo Combinado: GICC. [Internet]. 2006. Available from: http://biblioteca.uns.edu.pe/saladocentes/archivoz/curzoz/04_gt15_c_t_de_gasif_integrada_en_ciclo_combinado.pdf [Accessed: 2018-03-13]

[25] Mercer GE, Fitzgerald S, Day J, Filby RH. Determination of organic/inorganic associations of trace elements in kerogen of the New Albany shale. Fuel. 1993;**72**(11):1187-1195

[26] Vassilev SV. Trace elements in solid waste products from coal burning at some Bulgarian thermoelectric power station. Fuel. 1994;**73**(3):367

[27] Querol X, Fernandez TJL, Lopez SA. Trace elements in coal and their behavior during combustion in a large station. Fuel. 1995;**74**(3):331-343

[28] Querol X, Crehuet R, Alastuey A, López-Soler A, Fernández-Turiel JL, Plana F, Puicercus JA. Geoquímica y mineralogía aplicadas a estudios de impacto ambiental derivado de la combustión del carbón. Acta Geologica Hispanica. 1996;**30**(1-3):145-153

[29] Dale LS, Chapman JF, Lavrencic SA. In: Proceedings of the 7th International Conference on Coal Science; Alberta, Canada. 1993. pp. 313-316

[30] Shpirt MJ, Zeckel LA. Proceedings of the 7th International Conference on Coal Science. 1993. pp. 330-333

[31] Swaine DJ. The importance of trace elements in Australian coals. Energy News Journal. 1982;**4**(3):18-22

[32] Córdoba P, Ochoa-González R, Font O, Izquierdo M, Querol X, Leiva C, et al. Partitioning of trace inorganic elements in a coal-fired power plant equipped with a wet flue gas desulphurisation system. Fuel. 2012a;**92**:145-157

[33] Córdoba P, Font O, Izquierdo M, Querol X, Leiva C, López-Antón MA, Díaz-Somoano M, Martínez-Tarazona MR, Ochoa-González R, Gómez P. The retention capacity for trace elements by the flue gas desulphurisation system under operational conditions of a co-combustion power plant. Fuel. 2012;**102**:773-778

[34] Germani MS, Zoller WH. Vapour-phase concentrations of arsenic, selenium, bromine, iodine and mercury in the stack of a coal-fired power plant. Environmental Science and Technology. 1988;**22**(9):1079-1085

[35] Meij R. Tracking trace elements at a coal-fired power plant equipped with a wet flue-gas desulphurisation facility. Kema Scientific & Technical Reports. 7. NV Kema, Arnhem; 1989. [Internet]. Available from: https://www.researchgate.net/publication/281479670_Tracking_trace_elements_at_a_coal_fired_plant_equipped_with_a_wet_flue_gas_desulphurization_facility [Accessed: 2018-04-13]

[36] Goodarzi F. Assessment of elemental content of feed coal, combustion residues and stack emitted materials for a Canadian pulverized coal fired power plant, and their possible environmental effect. International Journal of Coal Geology. 2006;**65**:17-25

[37] Díaz-Somoano M, Unterberger S, Hein KRG. Prediction of trace element volatility during co-combustion processes. Fuel. 2006;**85**(7):1087-1093

[38] Góral D, Wylenżek A. The efficient coal alternative. Petroleum coke-fired CFB boilers in Europe. Presented at Coal Gen Europe Warsaw, Poland; February 14-16, 2012

[39] European Coal Combustion Products Association. [Internet]. Available from: www.ecoba.com [Accessed: 2018-06-22]

[40] Nugteren H. Secondary industrial minerals from coal fly ash and aluminium anodising waste solutions [thesis]. Delft: Technische Universiteit Delft; 2010

[41] Font O, Querol X, Huggins FE, Chimenos JM, Fernández AI, Burgos S, García-Peña F. Speciation of major and selected trace elements in IGCC fly ash. Fuel. 2005;**84**:1364-1371

[42] Aunela-Tapola L, Hatanpää E, Hoffren H, Laitinen T, Larjava K, Rasila P, Tolvanen M. A study of trace element behaviour in two modern coal-fired power plants: II. Trace element balances in two plants equipped with semi-dry flue gas desulphurization facilities. Fuel Processing Technology. 1998;**55**(1):13-34

[43] Heebink LV, Hassett DJ. Hg release from FGD. In: Proceeding of International Ash Utilization Symposium; CAER; University of Kentucky. 2003. p. 75

[44] Schroeder K, Kairies C. Distribution of Hg in FGD by-products. In: Proceedings of World of Coal Ash Conference; April 11-1; CAER; University of Kentucky. Paper 100

[45] Álvarez-Ayuso E, Querol X, Tomás A. Environmental impact of coal combustion desulphurisation plant: Abatement capacity of desulphurisation process and environmental characterisation of combustion by-products. Chemosphere. 2006;**665**:2009-2017

[46] Córdoba P, Font O, Izquierdo M, Querol X, Tobías A, López-Antón MA, et al. Enrichment of inorganic trace pollutants in re-circulated water streams from a wet limestone flue gas desulphurisation system in two coal power plants. Fuel Processing Technology. 2011;**92**:1764-1775

[47] López-Antón MA, Díaz-Somoano M, Fierro JLG, Martínez-Tarazona MR. Retention of arsenic and selenium compounds present in coal combustion and gasification flue gases using activated carbons. Fuel Processing Technology. 2007;**88**:799-805

[48] Noda N, Ito S. The release and behaviour of mercury, selenium, and boron in coal combustion. Powder Technology. 2008;**180**:227-231

[49] Córdoba P, Lieberman R, Izquierdo M, Moreno N, Font O, Quero X, Giménez A. Effect of the flue gas desulphurisation (FGD) operating conditions on the behaviour and speciation of inorganic trace pollutants in a co-firing power plant. Fuel. (undereview)

[50] Córdoba P. Partitioning and speciation of selenium in wet limestone flue gas desulphurisation (FGD) systems: A review. Fuel. 2017;**202**:184-195

[51] Galbreath KC, Zygarlicke CJ. Mercury transformation in coal combustion flue gas. Fuel Processing Technology. 2000;**65-66**:289-310

[52] Córdoba P, Maroto-Valer M, Ayora C, Perry R, Rallo M, Font O, Izquierdo M, Querol X. Unusual speciation and retention of Hg at a coal-fired power plant. Environmental Science and Technology. 2012;**46**:7890-7897

[53] Wo J, Zhang M, Cheng X, Zhong X, Jiang X, Xinhua X. Hg^{2+} reduction and reemission from simulated wet flue gas desulphurization liquors. Journal of Hazardous Materials. 2009;**172**:1106-1110

[54] Stergaršek A, Horvat M, Frkal P, Stergaršek J. Removal of Hg^0 from flue gases in wet FGD by catalytic oxidation with air—An experimental study. Fuel. 2010;**89**:3167-3177

[55] Omine N, Romero CE, Kikkawa H, Wu S, Eswaran S. Study of elemental mercury re-emission in simulated wet scrubber. Fuel. 2012;**91**:93-101

[56] Shao D, Hutchinson EJ, Cao H, Pan WP. Behaviour of chlorine during coal pyrolysis. Energy & Fuels. 1994;**8**:399-401

[57] Kikkawa H, Nakamoto T, Morishita M, Yamada K. New wet FGD process using granular limestone. Industrial and Engineering Chemistry Research. 2002;**41**:3028-3036

[58] Mori T, Matsuda S, Nakajima F, Nishimura T, Arika Y. Effect of Al^{3+} and F^- on the desulfurization reaction in limestone slurry scrubbing process. Industrial and Engineering Chemistry Process Design and Development. 1981;**20**:144

[59] Maroney P, Davidson SA, Katzberger SM, Cichazowicz JE. Particulate control integration with wet scrubber SO_2 removal and waste generation. In: Proceedings of the 6th Symposium on the Transfer and Utilization of Particulate Control Technology; Palo Alto, CA: Electric Power Research Institute; 1986. pp. 1-22

Chapter 9

Industrial Air Pollution Control

Amir Hossein Matin and Ali Reza Rahmati

Additional information is available at the end of the chapter

http://dx.doi.org/10.5772/intechopen.80678

Abstract

Ambient air contaminants have different adverse effects on human health, environment, and structures. Some pollutions are more toxic and have unfavorable effects on workers' and public health, for example, cyanide/isocyanide vapor produced in some processes or in burning of polyurethane compounds, which is a toxic gas that can kill or cause harms impossible to reverse. It is so necessary that air pollutants will be controlled and treatment will be provided for the workers and public who are exposed or exhausted to the environment. Industrial ventilation (general ventilation, dilution ventilation, and local exhaust ventilation) is an appropriate system to control indoor air pollutions. Local exhaust ventilation (LEV) has different segments such as hoods, fittings, collectors (air cleaners), stacks, and fans that could collect and treat indoor and outdoor air contaminants. Each well-designed segment of a local exhaust ventilation is a vital subject that can cause an appropriate or inappropriate performance of systems. A well-designed LEV can lead to obtain a high efficiency level of pollution removal and minimum exposure (workers, public, and environment) to pollutants and save costs and energy.

Keywords: air pollution, air pollution control, design, industrial ventilation, collectors

1. Introduction

The importance of clean air in the industrial work environment is well known. A progressive industry with its sophisticated operations utilizes an increasing number of chemical compounds of which many are highly toxic. It is possible that using such materials results in vapors, gases, particulates, and/or mists in the air of workplaces in concentrations that exceed safe levels. For example, heat stress can result in an unsafe work environment.

Effective and properly designed ventilation offers a solution to the problem of protecting workers. Ventilation can also serve to control moisture, odor, and other undesirable environmental materials.

The health hazard potential of an airborne substance is characterized by the threshold limit value (TLV). TLV refers to the airborne concentration of a substance. It represents the conditions that under which it is believed that nearly all workers may be exposed day after day without adverse health effects.

The ventilation systems used in industrial plants are of two kinds. The "supply" system is used to supply air to a work space. The "exhaust" system is used to remove the contaminants which are generated by an operation to maintain a healthful work environment. When a dilution ventilation system is used to control or isolate contaminants in a special area of the overall plant, this may be desirable. Often, this condition occurs simply because of installing local exhaust systems and not considering the corresponding replacement air systems.

A well-designed supply system will consist of an air inlet section, heating and/or cooling equipment, filters, a fan, and register/grilles and ducts for distributing the air within the work space. Exhaust ventilation systems are classified into three groups: (1) the "local" exhaust system. The general exhaust system can be used for controlling heat and/or removing nontoxic contaminants generated in a space by flushing out a given space with large quantities of air and (2) the "general" exhaust system. The air may be tempered and recycled when used for heat control. And when used for controlling contaminant (the dilution system), enough outdoor air must be mixed with the contaminant, so that the average concentration is reduced to a safe level. Then, the contaminated air is typically discharged to the atmosphere. A supply system is usually used in conjunction with a general exhaust system to replace the air exhausted. Local exhaust ventilation systems operate on the principle of capturing toxic contaminants at or near its source. It is the preferred method of control because it is more effective and compared to high flow rate general or dilution exhaust requirements, the smaller exhaust flow rate results in lower heating or cooling load costs. Dilution ventilation systems are normally used for controlling toxic contaminants when local exhaust is impractical or is not economic, as the large quantities of tempered replacement air required to offset the exhausted air can lead to high operating costs. The present emphasis on industrial air pollution control highlights the need for efficient air cleaning devices on industrial ventilation systems, and the smaller flow rates of the local exhaust systems result in lower costs for air cleaning devices and better efficiency. Local exhaust systems consist of four basic elements: The hood(s), the duct system, the air cleaning device, and the fan. The purpose of the hood is collecting the contaminant generated in an air stream directed toward the hood. Then, a duct system must transport the contaminated air to the air cleaning device (collectors) or to the fan. In the air cleaner, before the contaminant is exhausted to environment, it is removed from the air stream. While producing the intended flow rate, the fan must overcome all the losses due to friction of hood entry, ducts, and fittings in the system. Most of the time, the duct on the fan outlet discharges the air to the atmosphere in such a way that it will not be re-entrained by the replacement systems.

2. Local ventilation

Designing local exhaust systems aims to capture and remove process emissions prior to their escape into the workplace environment. The local exhaust hood is the point of entry into the exhaust system. Regardless of their physical configuration, it is defined herein to include all suction openings. The hood primarily is to create an air flow field which effectively captures the contaminant and transports it into the hood. In addition, local ventilation system contains four parts that its characteristics are calculated as following:

3. Hood flow rate

Hoods may be of different shapes and dimension configurations but can be categorized into two general groups, i.e., enclosing and exterior. The type of the hood to be used depends on the physical characteristics of the process equipment, the operator/equipment interface, and the contaminant generation mechanism. Enclosing hoods are those which partially or completely enclose the process or contaminant generation point. A complete enclosure may be a laboratory glove box. Wherever the process configuration and operation permit, enclosing hoods are preferred. Exterior hoods are those which are located adjacent to an emission source without enclosing it. Equation (1) indicates the calculation of flow rate of exterior hood by a general equation.

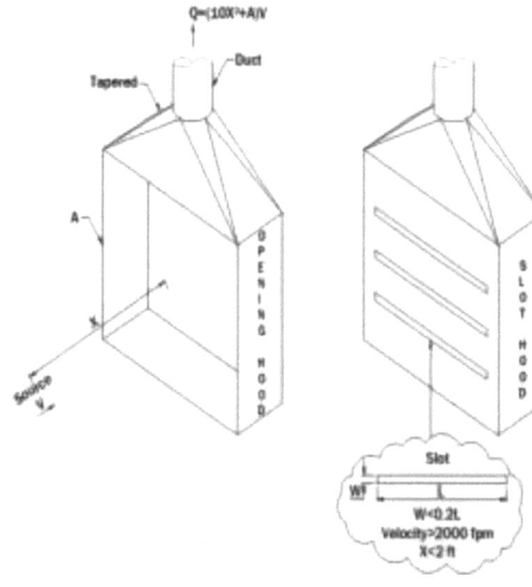

Calculation of flow rate of exterior hood: $Q = K_Q\,(10X2 + A)\,V$ (1)

where K_Q is the air correction factor, X is the pollution center to hood face (ft), A is the hood face area (ft^2), V is the capture velocity (fpm), and Q is the hood suction (cfm).

Calculation flow rate of standard VS of ACGIH for example. **Figure 1** shows calculation of flow rate of standard VS of ACGIH.

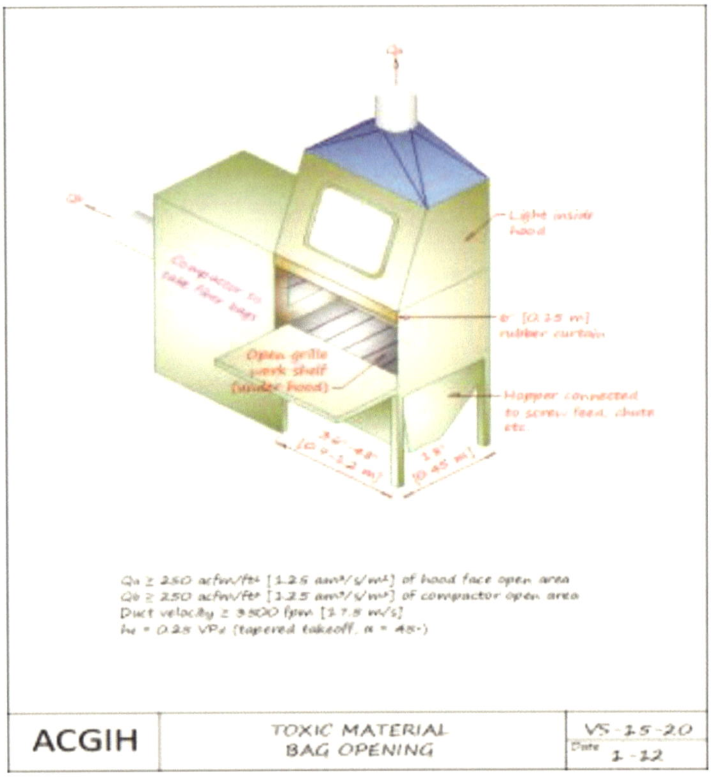

Figure 1. Calculation of flow rate of standard VS of ACGIH.

4. Ducts

Ducts are connected from hoods to fan and transfer air pollutions; therefore, the most velocity of air is in the duct. Then, the most losses are in the duct, and ducts show the static pressure of ventilation system. This process is more involved than merely connecting pieces of duct. Contaminant control may not be achieved if the system is not carefully designed in a manner which inherently ensures that the design flow rates will be realized. The results of the following design procedure will determine the material thickness, duct sizes, and the fan operating point required by the system.

All exhaust systems contain hoods, duct segments, and special fittings leading to an exhaust fan. A complex system is an arrangement of some simple exhaust systems which are connected to a common duct. There are two general types of duct system designs: tapered systems and plenum systems. The duct in a tapered system gradually gets larger since additional flows are merged together, thus keeping duct velocities nearly constant. The tapered system will maintain the minimum velocity required to prevent settling if the system transports particulate. The duct in a plenum system is generally larger than that in a tapered system, and the velocity in it is usually low. Any particulate in the air stream can settle out in the large ducts. Select or design each exhaust hood on the basis of the process, toxicity, and physical

http://dx.doi.org/10.5772/intechopen.80678

and chemical characteristics of the material and the ergonomics of the process, and then determine its minimum duct velocity, design flow rate, and entry losses. Note that minimum duct velocity is only important for systems transporting particulate, condensing vapors, or mist and for preventing explosive concentrations building up in the duct.

4.1. Duct segment calculation

The velocity pressure method is based on the fact that all frictional and dynamic (fitting) losses in ducts are functions of the velocity pressure and can be calculated by a loss factor multiplied by the velocity pressure. Loss factors are for straight ducts, elbows, and branch. Note that velocity pressure is always positive. Also, total pressure is always greater than static pressure when static and total pressures are negative at suction zoon and positive at air drift to atmosphere. Determine the hood static pressure. By the loss coefficient from the tabulated data, multiply the design duct length. Using galvanized sheet metal duct was assumed throughout this article. Determine the number and type of fittings in the duct segment. Add the results of and multiply by the duct VP. This is the actual loss in inches of water for the duct segment. Finally add the result to the hood suction. Add them in also if there are any additional losses (expressed in inches of water), such as for an air cleaning device. This establishes the cumulative energy required, expressed as static pressure, to move the design flow rate through the duct segment. It should be noted that the final value is negative. Equation (2) indicates the total pressure equation.

$$\text{Total pressure equation:SP (TP} = \text{SP} + \text{VP)} \tag{2}$$

where SP is the static pressure ("wg), VP is the velocity pressure ("wg), and TP is the total pressure ("wg).

The process of calculation of ventilation systems is as follows:

By the following formula, determine flow correction for air psychrometric and sea level elevation. By dividing the actual flow rate by the area of the commercial duct size chosen, determine the minimum duct design velocity, and then calculate actual velocity. Equation (3) shows the calculation of velocity pressure by the following formula:

$$\text{Velocity pressure equation:VP} = K_P \left(\frac{V}{4005}\right)^2 \tag{3}$$

where K_p is the pressure correction factor, VP is the velocity pressure (fpm), and V is the duct velocity ("wg).

4.2. Duct losses

Air movement has friction to the inside wall of the duct, so it creates loss that is calculated as follows:

4.2.1. Loeffler formula

Equation (4) shows the calculation of Loeffler, shown in **Table 1**, choices *a, b, and c*:

Duct materials	a	b	c
Galvanized-PVC-PE	0.0307	0.533	0.612
Iron-steel-aluminum	0.0425	0.465	0.602
Flexible duct wires covered	0.0311	0.604	0.639
Flexible duct wires exposed	0.0428	0.649	0.683

Table 1. Correlation equation constants

$$\text{Loeffler equation:} H_f = a\frac{V^b}{Q^c} \tag{4}$$

where H_f is the Loeffler coefficient; Q is the air flow rate (cfm); V is the duct velocity (fpm); l is the straight duct length (ft); VP is the duct velocity pressure ("wg); and a, b, and c are the coefficients of duct material.

5. Fittings

The fittings are the pieces that are mounted on the duct to redirect the path and branch. The standard fitting in industrial ventilation include:

5.1. Elbow

Elbows are used to redirect the air stream. In industrial ventilation, the standard elbows include 90°, 60° and 45°. It is practically impossible to curl the sheets, given that the elbow is made of metal sheets (black iron, galvanized, stainless steel, etc.); therefore, elbows must be made of pieces. Minimum segments in elbows 90° must be five segments, in elbows 60° must be four segments, and in elbows 45° must be three segments. **Figure 2** shows the standard elbows.

5.2. Radius of elbow

Distance center of elbow arc with longitudinal axis line in standard elbows in industrial ventilation has three types of circle radius. The elbow is as follows in the radius of rotation and its diameter: R = 1.5 d or R = 2.0 d or R = 2.5 d. The elbow R = 2.0 d is optimized for low-pressure drop and low turn radius, but R = 1.5 d is economical according to our experiences.

5.3. Elbow losses

The pressure drop in the elbows is calculated in Eq. (5), and **Table 2** indicates the elbow friction factor.

$$\text{Calculation of the pressure drop in the elbows: } h_f = K \cdot VP \tag{5}$$

http://dx.doi.org/10.5772/intechopen.80678

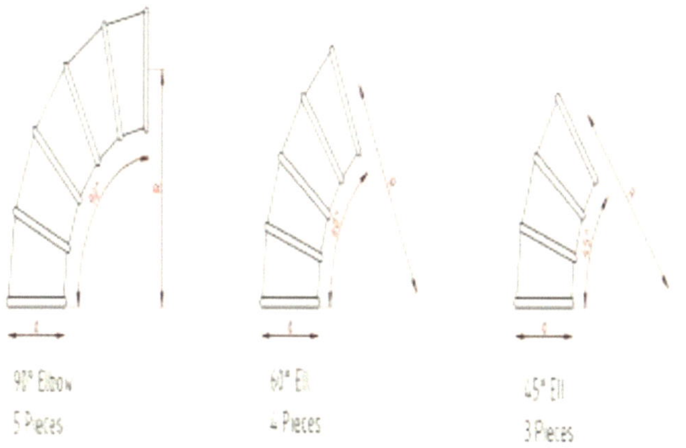

Figure 2. Standard elbows.

R/d	Friction factor
1.5	0.24
2.0	0.19
2.5	0.17

Table 2. Elbow friction factor

where h_f is the loss of elbow ("WG), K is the elbow friction factor ("WG), and VP is the velocity pressure ("WG).

5.4. Entrance

This is the branch of the flow of air flow from one duct to another duct distribution (see **Figure 3**) of air flow. Equation (6) shows the pressure drop in the entrance, and **Table 3** shows the entrance friction factor [1–3].

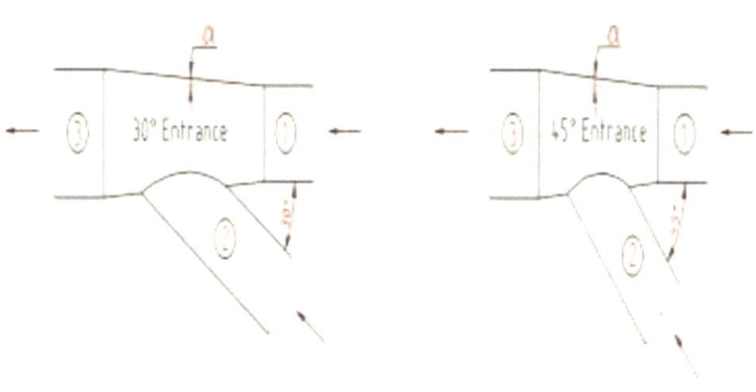

Figure 3. Entrance.

Θ degrees	Friction factor
30	0.18
45	0.28

Table 3. Entrance friction factor

$$\text{Pressure drop in the entrance equation}: h_f = K \cdot VP \tag{6}$$

where h_f is the loss of entrance ("WG), K is the entrance friction factor ("WG), and VP is the velocity pressure ("WG).

6. Air cleaning devices

Dusts, toxic or corrosive gases, and fumes should not be discharged to the atmosphere. Each exhaust system handling such materials should be provided with an adequate air cleaner. Air cleaning devices remove contaminants from an air or gas stream after ventilated from indoor spaces and before exhausted to the atmosphere. They are available in a wide range of designs to meet variations in air cleaning requirements. Quantity and characteristics of the contaminant to be removed, conditions of the air or gas stream, and degree of removal required will all have a bearing on the device selected for any given application. In addition, fire safety and explosion control must be considered in all selections. For particulate contaminants, air cleaning devices are divided into two groups: dust filters and air cleaners.

Air filters are designed to remove low dust concentrations of the magnitude found in atmospheric air. This kind of air cleaning device is typically used in air-conditioning, ventilation, and heating systems where dust concentrations seldom exceed 1.0 grains per thousand cubic feet of air and are usually well below 0.1 grains per thousand cubic feet of air. Where the air or gas to be cleaned originates in local exhaust systems or process stack gas effluents, usually duct collectors are designed for the much heavier loads from industrial processes. For each cubic foot of air or gas, contaminant concentrations will vary from less than 0.1 to 100 grains or more. Therefore, dust collectors are, and must be, capable of handling concentrations 100–20,000 times greater than those for which air filters are designed. Small, inexpensive versions of all categories of air cleaning devices are available. The principles of selection, application, and operation are the same as for larger equipment. However, much of the available equipment is of light duty design and construction due to the structure of the market that focuses on small, quickly available, and inexpensive equipment. One of the major economies of unit collectors implies recirculation, for which such equipment may or may not be suitable. Application engineering is just as essential for unit collectors as it is for major systems for adequate prevention of health hazards, fires, and explosions.

http://dx.doi.org/10.5772/intechopen.80678

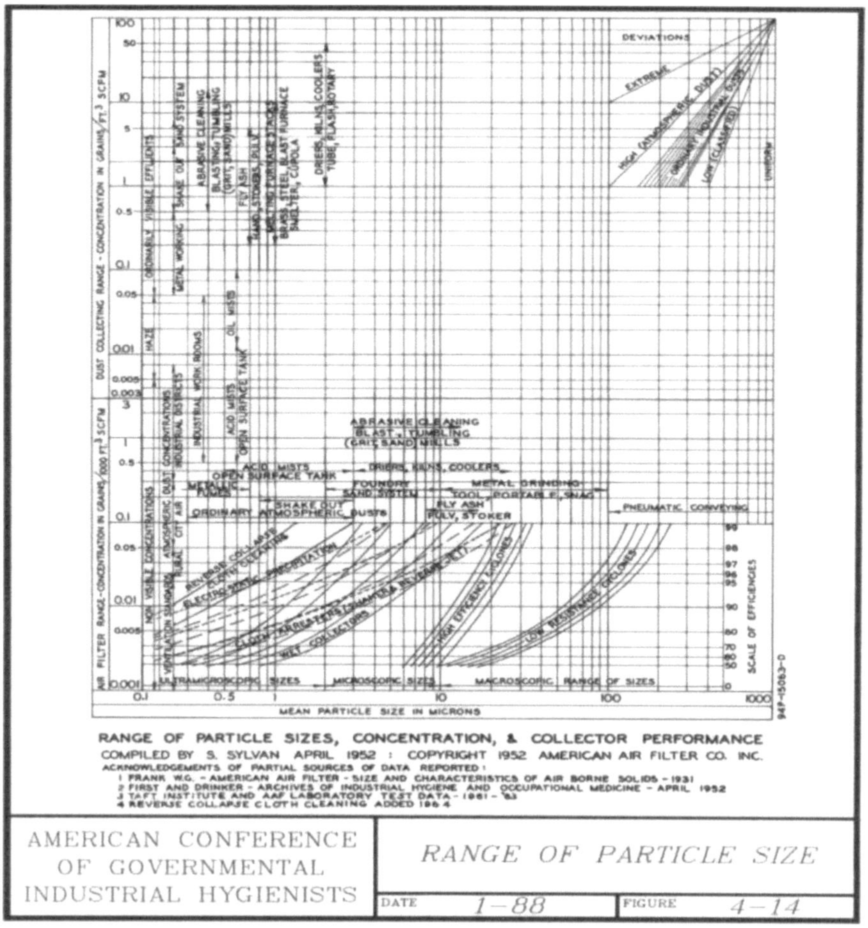

Figure 4. Selection of dust cleaners on the basis of range of particle size, concentration, and collector performance.

6.1. Selection of dust collection equipment

Contaminants in exhaust systems cover a wide range in concentration and particle size. Concentrations can range from less than 0.1 to much more than 100,000 grains of dust per cubic foot of air. The dust ranges from 0.5 to 100 or more microns in size in low-pressure conveying systems. Deviation from mean size (the range over and under the mean) will also vary with the material. **Figure 4** indicates the pattern of selecting dust cleaners on the basis of range of particle size, concentration, and collector performance.

7. Required efficiency

The required degree of collection can depend on plant location, nature of contaminant, and the regulations of governmental agencies when the cleaned air is to be discharged outdoors.

Damage to farms or contribution to air pollution problems of distant cities can affect the need for and importance of effective collection equipment in remote locations. Many industries, originally located away from residential areas, have failed to anticipate the residential building construction which frequently develops around a plant. Such lack of foresight has required installation of air cleaning equipment at greater expense. Nowadays, the remotely located plant must comply with the same regulations as the plant located in an urban area in most cases. Management can continue to expect criticism for excessive emissions of air contaminants whether located in a heavy industry section of a city or in an area closer to residential zones with the present emphasis on public nuisance, public health, and preservation and improvement of community air quality. Also, the mass rate of emission will affect selection of equipment. For a given concentration, the larger the exhaust volumetric flow rate, the greater the need for better equipment. While a smaller industrial pulverized fuel boiler might be able to use slightly less efficient collectors, large central steam-generating stations might select high efficiency electrostatic precipitators or fabric collectors for their pulverized coal boiler stacks.

A safe recommendation in selecting equipment is to select the collector that will allow the least possible amount of contaminant to escape and also while meeting all prevailing air pollution regulations is reasonable in first cost and maintenance. It must be remembered that visibility of an effluent will be a function of the light reflecting surface area of the escaping material. Surface area per pound increases inversely as the square of particle size. In other words, the removal of 80% or more of the dust on a weight basis may remove only the coarse particles without altering the stack appearance.

The contaminant characteristics will also affect equipment selection. Emitted chemicals may attack collector elements or corrode wet type collectors. Sticky materials can adhere to collector elements, plugging collector passages. Linty materials adhere to certain types of collector surfaces or elements. Abrasive materials in moderate to heavy concentrations will cause rapid wear on dry metal surfaces. Particle size, density, and shape will rule out certain designs. The combustible nature of many finely divided materials requires specific collector designs to assure safe operation. The characteristics of the carrier gas stream can have a marked bearing on selecting equipment. It is possible that temperature of the gas stream limit the material choices in fabric collectors. Condensation of water vapor will cause packing and plugging of air or dust passages in dry collectors. We can reach optimum and high removal efficiency with optimization and more study about design parameters in each type of collectors.

8. Dust collector types

There are four major types of dust collectors for particulate contaminants: electrostatic precipitators, fabric collectors, wet collectors, and dry centrifugal collectors.

8.1. Fabric collectors

The "fabric" may be constructed of any fibrous material, either man-made or natural, and regardless of construction may be spun into a yarn and woven or felted by needling, impacting, or bonding. The fabric represents a porous mass through which the gas is passed in directional such that dust particles are retained on the dirty side and the cleaned gas passes on through. The ability of the fabric to pass air is called "permeability." It is defined as the cubic feet of air passed through one square foot of fabric each minute at a pressure drop of 0.5 ″wg. Usual permeability amounts for commonly used fabrics range from 25 to 40 cfm. A highly efficient fabric that cannot be cleaned represents an excessive resistance to air flow and is not an economical engineering solution. Final fabric selection is generally a compromise between efficiency and permeability. The efficiency of the fabric as a filter is meaningful when new fabric is first put into service. Even after cleaning, a residual and/or redeposited dust cake provides higher collection efficiency and additional filtration surface than obtainable with new fabric. Fabric collectors are not 100% efficient. But well-designed, adequately sized, and properly operated fabric collectors can be expected to operate at efficiencies in excess of 99% and often as high as 99.9% or more on a mass basis. The fabric collector should be leak tested for mechanical leaks where extremely high collection efficiency is essential. The combination of fabric and collected dust becomes increasingly efficient as the dust cake accumulates on the fabric surface. Fabric collectors are suitable for service on relatively heavy dust concentrations. The amount of dust collected on a well-designed and single square yard of fabric may exceed 5 pounds per hour. Commercially available fabric collectors employ fabric configured as bags or tubes, envelopes (flat bags), rigid elements, or pleated cartridges. Most of the available fabrics are employed in either bag or envelope configuration. The variable design features of the many available fabric collectors are as follows:

1. Housing configuration (single compartment and multiple compartment)
2. Type of reconditioning (shaker, reverse air, pulse-jet)
3. Fabric configuration (bags or tubes, cartridges, envelopes)
4. Type of fabric (woven or non-woven)
5. Intermittent or continuous service

Due to many variables and their range of variation, fabric collector sizing is judged based on experience. Also, a combination of shaking and reverse air flow has been utilized. It is possible that reverse-jet, continuous-duty fabric collectors use envelopes or tubes of non-woven fabric, pleated cartridges of non-woven mat (paper-like) in cylindrical or panel configuration, or rigid elements such as sintered polyethylene. Based on our experience, when tubes have 6–11 inches diameter and can be as long as 10 feet, permeability 10–25, reverse-jet 6–8 atmosphere (in high load of pollution each 1 minute, 1 second pulse jet; and in low load of pollution each 2–3 minute, 1 second pulse jet), and air velocity inside the chamber selected 300 fpm, the bag filter becomes optimal and economic in removal efficiency. Solenoid valves which control the pulses of compressed air may be open for a tenth of a second or less. An EPA-sponsored research has shown that superior performance results from downward flow of the dirty air stream. This downward air flow reduces redeposition since it aids gravity in moving dust particles toward the hopper. **Figure 5** shows the fabric collector [1, 2].

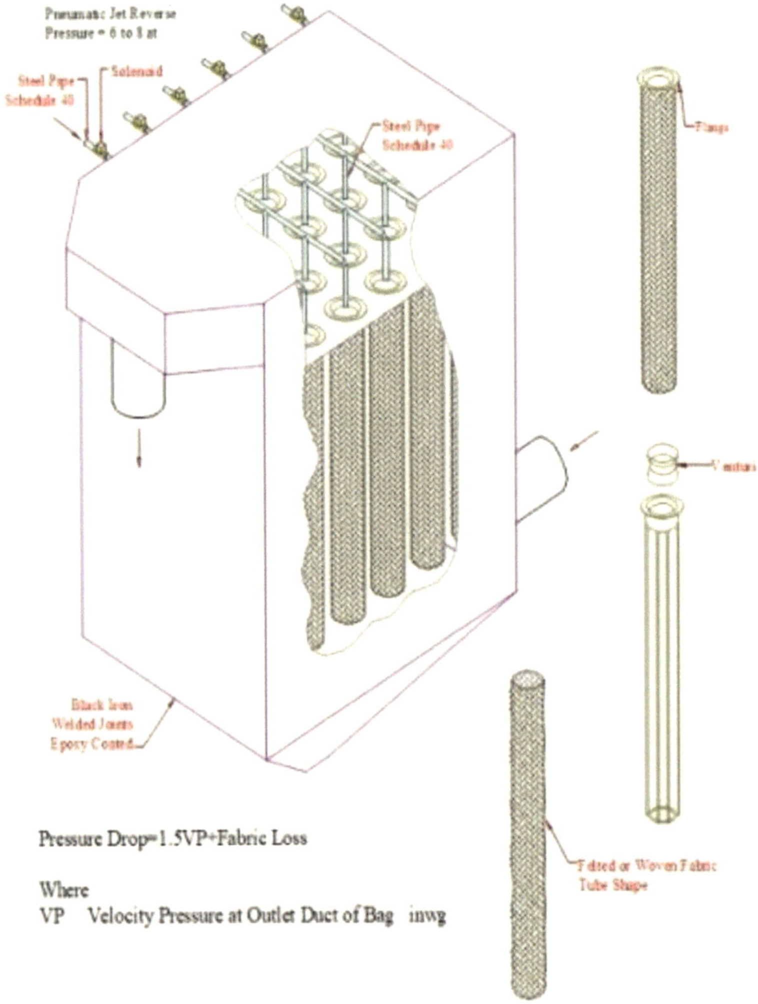

Figure 5. Fabric collector.

8.2. Wet collectors

Wet collectors or scrubbers are commercially available in different designs, with pressure drops from 1.5 times of exit duct velocity pressure. There is a corresponding variation in collector performance. It is generally accepted that efficiency depends on the energy utilized in air to water contact and is independent of operating principle for well-designed equipment. Whether the energy is supplied to the air or to the water, efficiency is a function of total energy input per cfm. In other words, if equivalent power is utilized, well-designed collectors by different manufacturers provide similar efficiency. Wet collectors have the ability to handle high-temperature and moisture-laden gases. The collection of dust in a wetted form minimizes a secondary dust problem in disposal of collected material. Some dusts represent explosion or fire hazards when it is dry. Wet collection minimizes the hazard; however, the use of water may introduce corrosive conditions within the collector, and if collectors are located outdoors in cold climates, freeze protection may be necessary. Wet collectors are frequently the solution to air pollution problems. It should be recognized that disposal of collected material in water without clarification or treatment may create water pollution problems. Wet collectors have one characteristic not found in other collectors, the inherent ability to humidify. Humidification, the process of adding water vapor to the air stream through evaporation, may be either advantageous or disadvantageous depending on the situation. All wet collectors humidify; the amount of humidification varies for different designs. According to our experiences, although wet collectors have different types, the spray tower and packed towers, which are more practical and economic, are simple to make and assemble and have appropriated removal efficiency in air pollution control.

8.2.1. Spray tower

Spray tower collectors consist of a rectangular or round chamber into which water is introduced by spray nozzles. There are many variations of design, but the principal mechanism is impaction of dust particles on the liquid droplets created by the nozzles. These droplets are separated from the air stream by pump force. The pressure drop is relatively low (on the order of 0.5–1.5 ″wg), but water pressures range from 10 to 400 psig. The high-pressure devices are the exception rather than the rule. This type of collector generally utilizes low-pressure supply water and operates in the lower efficiency range for wet collectors. Collection efficiency can reach the upper range of wet collector performance where water is supplied under high pressure. For conventional equipment, water requirements are reasonable with a maximum of about 5 gpm per thousand cfm of particle, and fogging types using high water pressure may require as much as 10 gpm per thousand cfm of gas. **Figure 6** shows this collector. Air flow inside all of the scrubbers has adiabatic revolution with 90% efficiency; in other words, air flow in scrubber will be cool and increase humidity to 90% relative humidity. **Figure 7** shows this change.

Based on our experiences, design velocity in spray towers takes 250 feet per minute (fpm), and the performance of spray tower will be optimum since no droplet in fan housing or stack was detected. Design parameters of spray towers can influence the removal efficiency of air pollutants. Design parameters included type (solid cone or hollow cone), array and position and number of spray nozzles, size, liquid pressure, diffraction angle of spray nozzle, L/G (liquid to gas ratio), etc. The operating pressure of scrubbing liquid not only is an important parameter but also directly influences the liquid distribution, droplet size, and liquid flow rate. According to our experience, operating pressure, nozzle size 80–800 micron (the nozzle size directly depends on Henry's law constant of each gas), and multistage spray nozzle could increase mass transfer and removal efficiency, decrease and save operational costs, and optimize the "mechanical" performance of spray towers. The area and pressure drop of spray towers were calculated Eq. (7) [2, 4–6]:

$$\text{The area and pressure drop of spray towers:}\quad \begin{aligned} d_s &= 2\sqrt{\frac{Q}{14}} \\ b &= 2.5\, d_s \\ h_{1S} &= 1.5 \times VP \end{aligned} \tag{7}$$

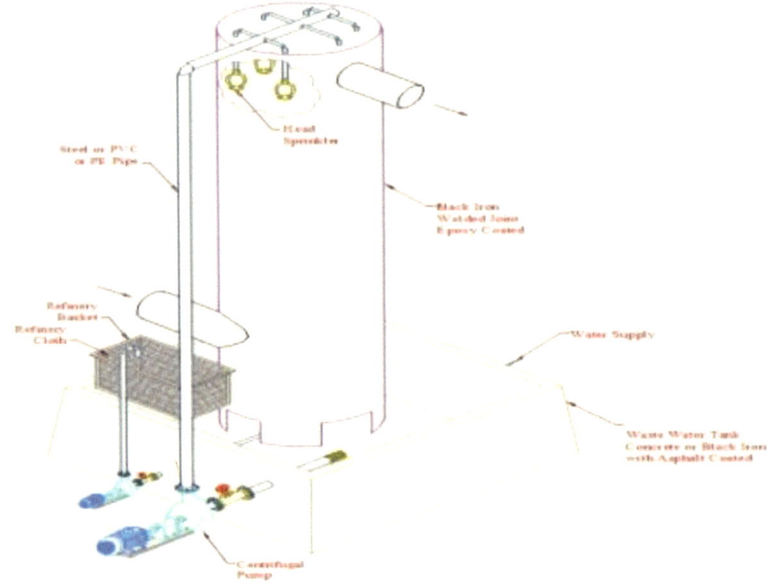

Figure 6. Spray tower.

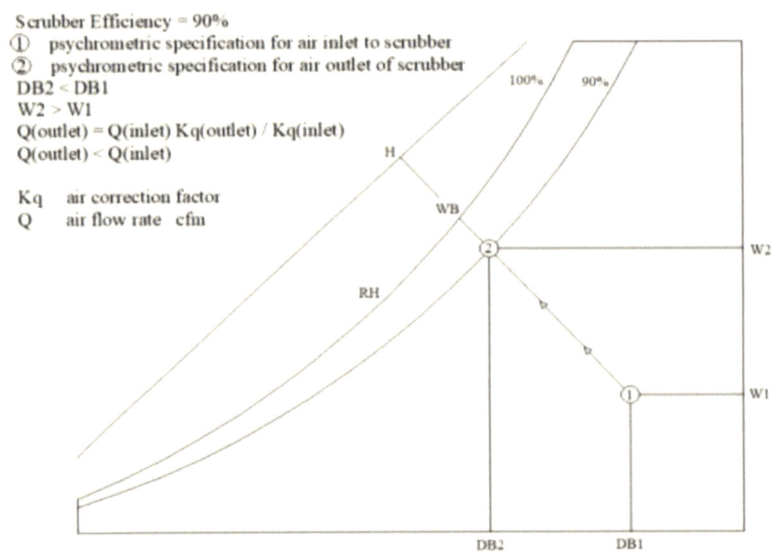

Figure 7. Adiabatic revolution.

where Q is the flow rate at scrubber outlet (cfm), b_S is the inlet duct to nozzle height (ft), h_{IS} is the scrubber pressure drop ("WG), d_s is the scrubber diameter (ft), and VP is the velocity pressure at the outlet duct ("WG).

http://dx.doi.org/10.5772/intechopen.80678

8.3. Packed towers

Packed towers are essentially contact beds through which liquid and gases pass concurrently, counter-currently, or in cross-flow. They are used primarily for applications involving vapor, gas, and mist removal. These collectors can capture solid particulate matter, but they are not used for that purpose since dust plugs the packing and requires unreasonable maintenance.

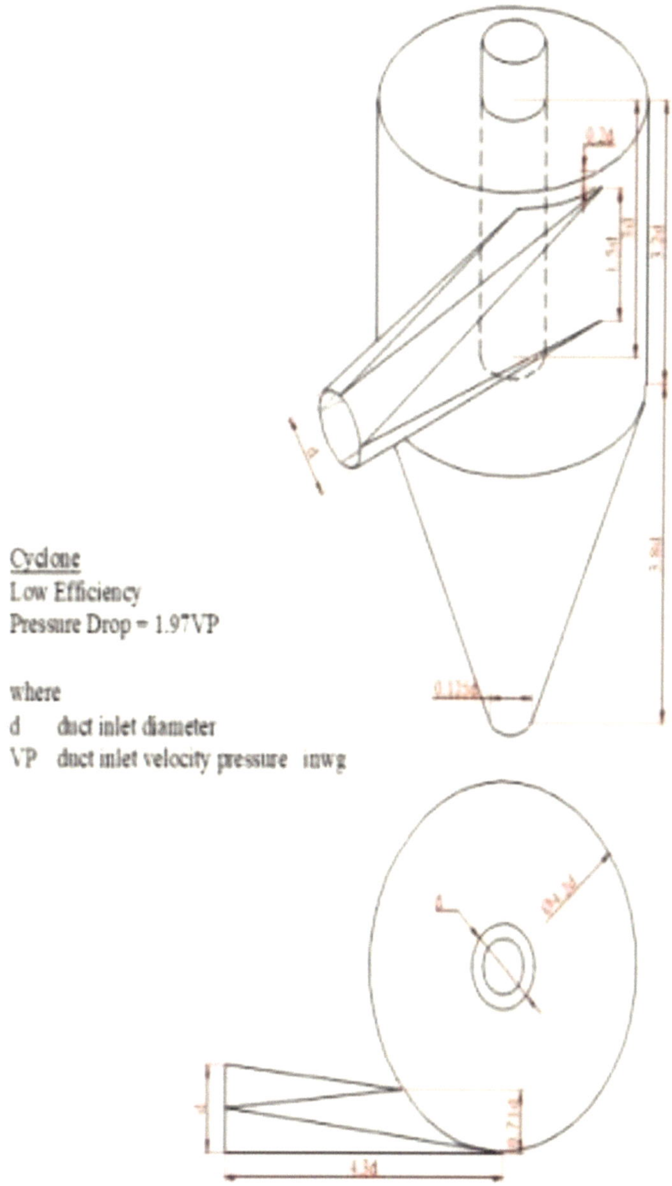

Figure 8. Low-efficiency cyclone.

Water rates of 5–10 gpm per thousand cfm are typical for packed towers. Water is distributed over V-notched ceramic or plastic weirs. High-temperature deterioration is avoided by using brick linings, allowing gas temperatures as high as 1600 F to be handled directly from furnace flues. Based on shapes, the packing is divided into the following types:

Maspac, Intalox Saddle, Pall Ring, Tellerette, Raschig ring, and Berl Saddle. On the basis of our experience, the Raschig ring is appropriate for packing (popular, simple to make and maintenance, low-cost, etc.), and the optimum design velocity was 250 fpm for packed tower.

8.4. Cyclone collector

The cyclone collector is commonly used for the removal of coarse dust from an air stream as a precleaner to more efficient dust collectors and/or as a product separator in air conveying systems. Principal advantages are low-cost, low maintenance, simple making, and relatively low-pressure drops. It is not suitable for the collection of fine particles.

8.4.1. Low-efficiency cyclone

In this collector, air velocity is 3200–4000 fpm with pressure drops 0.75–1.5 ″wg. This cyclone can absorb with efficiency more than $90^0/_0$ the particle size more than 55 microns (see **Figure 8**).

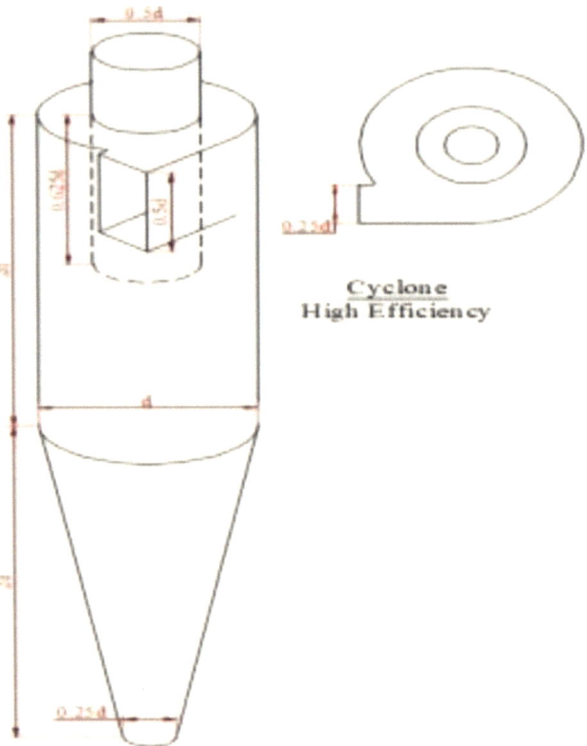

Figure 9. High-efficiency cyclone.

8.4.2. High-efficiency cyclone

In this collector, air velocity is 4000–4800 fpm with pressure drops 3–6″wg. This cyclone (see **Figure 9**) can absorb with efficiency more than $90^{0}/_{0}$ the particle size more than 13 micron. The diameter of this collector can be calculated based on Eq. (8):

$$\text{The diameter of high efficiency cyclone:d} = \left(\frac{0.083}{\text{hl}}\right)^{0.25}\sqrt{\text{Q}},\ \text{hl} = 3\ \text{to}\ 6\ (''\ \text{WG}) \qquad (8)$$

where Q is the cyclone flow rate (cfm), hl is the cyclone pressure drop (″WG), and d is the cyclone diameter (in).

8.4.3. Multiple cyclone

These collectors (see **Figure 10**) consist of a chamber that some number of high-efficiency cyclone put on this chamber. Not only the inlets of cyclones are connected together but also

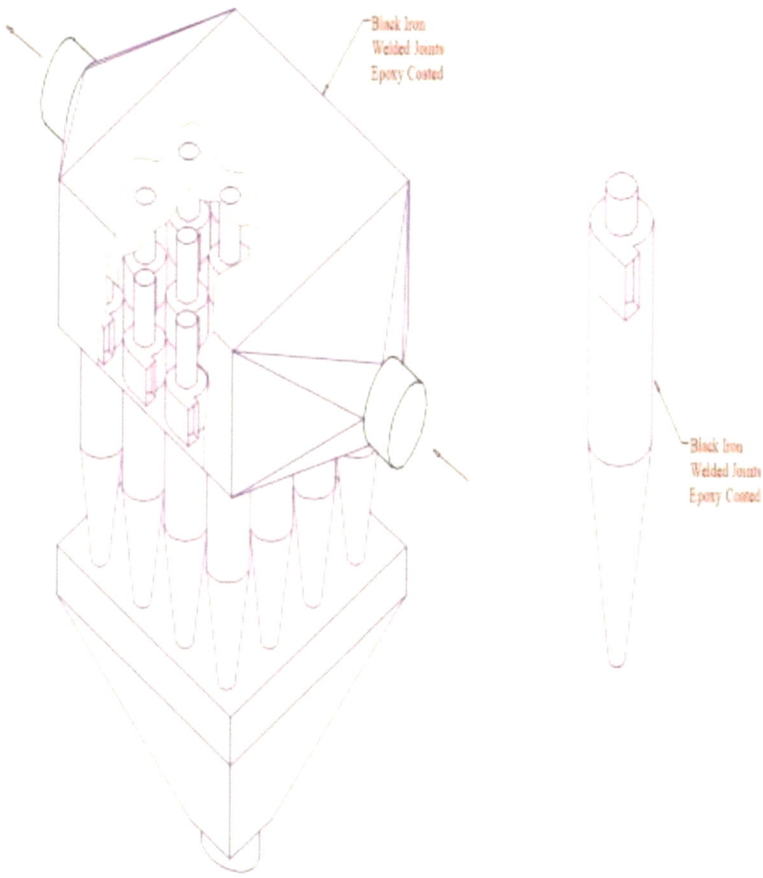

Figure 10. Multiple cyclone.

the outlets are connected together. This collector at list consist of four cyclones, and can more number cyclone put in this chamber and cyclones must have square array. The pressure drop in multi-cyclone is equivalent to each cyclone. The flow rate in each cyclone can be calculated based on Eq. (9) [1, 2]:

$$\text{Flow rate in each cyclone:} Q_i = \frac{Q}{N} \quad N = n^2 \tag{9}$$

where Q is the quota air for each cyclone, *Q* is the inlet or outlet air in cyclone (cfm), *N* is the number of cyclone in multi-cyclone, and n is the inlet eger number.

9. Conclusions

Local exhaust ventilation is the best option for removing and mitigating all industrial air pollutants, which reduces the cost of energy and the occupational and environmental exposure of individuals with a variety of environmental pollutants. Industrial ventilation is strongly recommended for environmentally polluting industries and workplaces.

Conflict of interest

The authors are able to design and construct all segments of the ventilation systems such as fans, ducts, fittings, and collectors with great knowledge and experience.

Acronyms and abbreviations

VS	ventilation standard
ACGIH	American Conference of Governmental Industrial Hygienist

Author details

Amir Hossein Matin[1] and Ali Reza Rahmati[2]*

*Address all correspondence to: alrahmati@aol.com

1 School of Health, Safety and Environment, Shahid Beheshti University of Medical Sciences, Tehran, Iran

2 School of Public Health, Shahid Beheshti University of Medical Sciences, Tehran, Iran

References

[1] Industrial Ventilation: A Manual of Recommended Practice For Design. In: ACGIH. 27th ed. 2013

[2] Matin AH. Industrial Ventilation Design and Calculation Guide for Industrial Hygienists. pp. 1-315

[3] Matin AH, Bayatian M. Collection of Air Pollutants. Fanavaran; 1389

[4] Jafari MJ, Matin AH, Rahmati AR, Azari MR, Omidi L, Hosseini SS, et al. Experimental optimization of a spray tower for ammonia removal. Journal of Atmospheric Pollution Research. 2018

[5] Rahmati AR. The influences of spray nozzles and water head on removal efficiency of ammonia from air in a spray tower [MSc thesis]. Tehran, Iran: School of Health, Shahid Beheshti University of Medical Sciences; 2015

[6] Jafari MJ, Nourian S, Zendehdel R, Massoudinejad MR, Sarbakhsh P, Rahmati AR, et al. The performance of a spray tower in scrubbing H_2S from air. Safety Promotion and Injury Prevention. 2015;**4**(2):321-328